ALTERNATIVE FUELS—
THE FUTURE OF HYDROGEN

ALTERNATIVE FUELS—
THE FUTURE OF HYDROGEN

BY MICHAEL FRANK HORDESKI

THE FAIRMONT PRESS, INC.

CRC Press
Taylor & Francis Group

Library of Congress Cataloging-in-Publication Data

Hordeski, Michael F.
    Alternative fuels : the future of hydrogen / by Michael Frank Hordeski.
      p. cm.
    Includes bibliographical references and index.
    ISBN 0-88173-519-1 (Fairmont Press, Inc.) -- ISBN 0-88173-520-5 (ebook)
    1. Hydrogen as fuel. 2. Hydrogen cars. 3. Hydrogen as fuel--Economic
aspects. 4. Energy policy--United States. I. Title.

    TP359.H8H67 2006
    333.79′4--dc22

                                       2006041264

Published by The Fairmont Press, Inc.
700 Indian Trail
Lilburn, GA 30047
tel: 770-925-9388; fax: 770-381-9865
http://www.fairmontpress.com

Distributed by Taylor & Francis Ltd.
6000 Broken Sound Parkway NW, Suite 300
Boca Raton, FL 33487, USA
E-mail: orders@crcpress.com

Distributed by Taylor & Francis Ltd.
23-25 Blades Court
Deodar Road
London SW15 2NU, UK
E-mail: uk.tandf@thomsonpublishingservices.co.uk

Printed in the United States of America
10 9 8 7 6 5 4 3 2 1

0-88173-519-1 (The Fairmont Press, Inc.)
0-8493-8236-X (Taylor & Francis Ltd.)

While every effort is made to provide dependable information, the publisher,
authors, and editors cannot be held responsible for any errors or omissions.

TABLE OF CONTENTS

PREFACE

Oil prices over $60 a barrel, soaring Chinese demand, rocketing energy markets, climate-destabilizing carbon emissions, new energy investments at $500 billion/year, the energy world has lost its bearings. Not since the energy shocks of the 1970's has the availability of energy been so important.

A recent survey indicated Americans believe energy security should be a top priority of U.S. energy policy with wide support for a moon shot type of effort to develop a hydrogen economy. The dependence of the U.S. on oil creates a national security vulnerability that could result in widespread economic problems and increased global instabilities.

Many factors affect our energy use, one of the most important is the availability of fuels. Mineral fuels can be divided into three types: solid, liquid and gas. In the first group are the coals. In the second group are the petroleum products which are rich in both carbon and hydrogen. These products provide a large range of fuels and lubricants. In the third group are the natural gases from petroleum deposits, the butane gases and coal and coke gas. Liquids include gasolines (or petrols). Their physical state allows them to be used directly in spark-ignition engines.

The two principal combustible elements common in coal and petroleum are carbon and hydrogen. Of the two, hydrogen is more efficient. The value of a fuel depends mainly in its calorific value. Pure carbon has a calorific value of 14,137 Btus while hydrogen has a value of 61,493 Btus. The higher the proportion of hydrogen a fuel contains, the more energy it will provide. The hydrogen content of liquid and gaseous fuels ranges from 10 to 50% by weight. They provide far more heat than solid fuels, which range from 5% to 12% by weight. The less oxygen in the fuel, the more easily the hydrogen and carbon will burn. The lower the oxygen content of a fuel, the better it will burn. The ideal fuel would be pure hydrogen.

Other factors used in assessing the merits of different fuels include moisture content, extraction, storage and transportation. Most fuels come directly or indirectly from carbohydrates; vegetable matter which result from photosynthesis occurring in green plants. The energy in these fuels is due to the sun. When burning fuels are extracted from living plants, we are recovering recent solar energy. When burning coal or gas, we are tapping ancient solar energy.

Is a hydrogen economy a reality? When President George Bush proposed a $1.7 billion program to promote hydrogen-fueled cars in his State of the Union Address, both sides of the aisle applauded. Almost everyone supported a hydrogen economy.

Hydrogen is the most abundant element on the planet. But it cannot be harvested directly. It must be extracted from another material. A wide variety of materials contain hydrogen, which is one reason it has attracted widespread support.

Environmentalists envision an energy economy where hydrogen comes from water, and the energy used to accomplish this comes from wind. The nuclear industry sees a water-based hydrogen economy, but with nuclear as the power source that electrolyzes water. *Nucleonics Week* views nuclear power as the only way to produce hydrogen on a large scale without adding to greenhouse gas emissions.

In the fossil fuel industry, they see hydrocarbons as providing most of our future hydrogen. They already have a head start since almost 50% of the world's commercial hydrogen now comes from natural gas and another 20% is derived from coal.

The automobile and oil companies are betting that petroleum will be the hydrogen source of the future. It was General Motors, that coined the phrase "the hydrogen economy".

A hydrogen economy will not be a renewable energy economy. For the next 20-50 years hydrogen will overwhelmingly be derived from fossil fuels or with nuclear energy.

It has taken more than 30 years for the renewable energy industry to capture 1% of the transportation fuel market (ethanol) and 2% of the electricity market (wind, solar, biomass).

A hydrogen economy will require the expenditure of hundreds of billions of dollars to build an entirely new energy infrastructure (pipelines, fueling stations, automobile engines). This will come from public and private money.

Making hydrogen takes energy. We may use a fuel that could be used directly to provide electricity or mechanical power or heat to instead make hydrogen, which is then used to make electricity.

We can run vehicles on natural gas or generate electricity using natural gas now. Converting natural gas into hydrogen and then hydrogen into electricity increases the amount of greenhouse gases emitted.

Hydrogen is the lightest element, being about eight times lighter than methane. Compacting it for storage or transport is expensive and energy intensive.

Another rationale for making hydrogen is that it is a way to store energy. That could benefit renewable energy sources like wind and sunlight that cannot generate energy on demand.

Many see hybrid vehicles as a bridge to a new type of transportation. Toyota and Honda have been selling tens of thousands of cars that have small gas engines and batteries. American automobile companies are following their lead.

Toyota and Honda and others are looking in the future to substitute a hydrogen fuel cell for the gasoline engine. That work will continue.

This book will get beyond the glib, "we can run our cars on water," news bites and assesses the reality to convert to a hydrogen economy. Some believe that the hydrogen economy has serious, perhaps fatal shortcomings while others consider it the path to a future of relative energy independence.

Topics include energy policy, fuel supply trends, statistics and projections, oil reserves, alternative fuel scenarios, fuel utilization, sustainable energy paths, cost analysis, fuels and development and regulatory issues. An energy evolution is taking place and will be pushed by changes in the environment, alternative fuels, sources of hydrogen, power generation needs, fuel cell/electric vehicles and ultimately provide our power and transportation future.

Chapter one is an overview of the barriers to implementation. It introduces the various technology and air emission issues, safety, and alternatives such as natural gas, hydrogen gas, methanol, ethanol and fuel cells.

Chapter two investigates the evolution of oil supplies, coke and gas fuels, city gas, natural gas, petroleum and sulphur, sources, crude oil prices, refining and distribution.

Fuels and autos are topics of Chapter three. It includes the auto future, electric cars, revivals, the auto industry, car designs and the impact of mass production.

Chapter four investigates the impact of auto technology. It considers fuel cell electric cars, fuel cell cabs, the fuel cell future and recent advances in fuel cell auto technology.

The new transportation is the theme of chapter five. Fueling stations are important for any alternative fuel especially hydrogen. Fuel cell advances will also pace hydrogen cars as well as the level of government support. An interesting concept for hydrogen cars is their proposed use as mobile utilities.

Chapter six outlines environmental trends and concerns. This includes, Kyoto and global warming, temperature cycles, deforestation and the greenhouse effect.

Chapter seven is concerned with hydrogen production and storage choices. Biomass is considered as a source of hydrogen fuel.

Alternative fuel programs is the theme of Chapter eight. Iceland's hydrogen plans are reviewed. The chapter ends with a discussion of solar hydrogen.

Hydrogen infrastructure is the leading issue at the beginning of Chapter nine. The book concludes with a discussion of nuclear hydrogen.

Many thanks to Dee who did much to organize the text and get this book in its final shape.

CHAPTER 1

FUELS AND TRENDS

In Britain, a successful hydrogen experiment was financed in the town of Harnosand by the Swedish steel industry, SAAB and other firms. In a house designed and lived in by Olaf Tegstrom, electric power was provided by a small computer-controlled Danish windmill in the garden. The power was used to electrolyse filtered water into hydrogen and oxygen. The hydrogen gas was used for cooking and heating the house and as fuel for a SAAB car. The car is almost non-polluting since the exhaust consists mostly of water vapor which is safe to drink. The hydrogen gas is absorbed in a metal hydrid and released as required.

In West Berlin, thanks to government subsidies for fuels that do not cause acid rain, Daimler Benz has built a filling station where converted vehicles can be filled with hydrogen, produced from town gas.

If hydrogen could become the prime provider of energy, a technological revolution could take place that would solve the problems of atmospheric pollution and oil depletion. Hydrogen has an energy content three to four times higher than oil, and it can be produced from all known energy sources, besides being a by-product of many industrial processes.

Hydrogen powered fuel cells could have wide applications, replacing batteries in many portable application, vehicle and using hydrogen for home electrical needs.

THE FUTURE FOR ALTERNATE FUELS

As oil prices increase, the interest in alternative fuels increases. This is evidenced by demonstration programs and commitments by states such as California. The concern of the air quality in many areas around the world makes finding solutions more urgent. As the price of oil rises, alternate fuels become more competitive. Major questions remain to be answered on which fuel or fuels will emerge and to what extent alternative sources will replace gasoline as the main product of crude oil.

1

Although it may be difficult, more cooperation is required between vehicle manufacturers, fuel producers, and the government. The infrastructure for the production and delivery of the fuels can evolve as needed with free market forces providing most of the momentum. But, there will need to be a coordination of selections of fuels and the adjustments needed to run those fuels.

A combination of available alternative fuels will evolve with the most likely choices affected by a number of technical, political and market factors. In order to allow a wider application of alternative fuels, a number of obstacles have to be overcome. These include economic, technological, and infrastructural issues. In the past, gasoline has been plentiful and has had a significant price advantage compared to other fuels. This could change quickly and alternative fuels would need to become more commonplace. One of the alternatives involves the more widespread use of biomass produced fuels. More efficient biomass conversion techniques would help make biofuels more cost-competitive.

FUEL FROM BIOMASS

Land availability and crop selection are major issues in biomass fuel usage. Biomass alternatives are expected to grow to a significantly larger scale for providing fuel. Land availability may not be a major problem, but land use issues need to be coordinated. Long-term production of biofuels in substantial quantities will require a number of changes. Present grain surpluses may not provide sufficient feedstocks for the fuel quantities needed. Producers may need to switch to short-rotation woody plants and herbacous grasses, feedstocks that can sustain biofuel production in long-term, substantial quantities. The increased use of municipal solid waste (MSW) as a feedstock for renewable fuels is likely to grow.

In spite of significant problems, many are optimistic about the role of biomass for alternative fuels in the future. The U.S. Department of Energy believes that biofuels from nonfood crops and MSW could potentially cut U.S. oil imports by 15 to 20%. Ethanol industry members believe that the capacity for producing that fuel alone could be doubled in a few years and tripled in five years.

Methanol

Methanol, which is also known as wood alcohol, is a colorless and odorless liquid alcohol fuel that can be made from biomass, natural gas, or coal. It is the simplest alcohol chemically and it may be used as an automobile fuel in its pure form (M100), as a gasoline blend typically 85% methane to 15% unleaded gasoline (M85) or as a feedstock for reformulated gasoline, or methyl tertiary butyl ether (MTBE). M100 or pure methanol is used as a substitute for diesel. In M85, the gasoline is added to color the flame of burning fuel for safety reasons and to improve starting in cold weather.

In order to achieve more methanol utilization, production needs to become more efficient and the infrastructure improved to make it more competitive. A major source of methane has been natural gas, since this has been the most economical source. Although the United States has vast quantities of both natural gas and coal, these are both nonrenewable resources.

Biomass can be a renewable feedstock for methane. Biomass feedstocks for methane production include crop residues, municipal solid waste (MSW), and wood resources. Biomass resources for the production of alcohol fuels are estimated at about 5 million dry tons per day which could provide 500 million gallons of methanol per day.

By 1991 the U.S. methanol industry was producing almost 4 million gallons of methanol per day. About a third of this was used as fuel for transportation and much of this methanol was converted to MTBE. Methanol is also popular in high-performance racing because of its octane-enhancing qualities.

In California there were more than 1,000 methanol vehicles including cars, trucks, and buses on the road in a state program with auto manufacturers and oil companies. In 1992, New York City also had a number of buses that ran on methanol. Arizona Checker Leasing purchased its first methanol vehicle in 1993 and now has 300 in its fleet of 450 M85 fuel flexible vehicles.

Ethanol

Ethanol, or grain alcohol is an alcohol fuel that has been more widely used as automotive fuel. It can be made from a variety of feedstocks, mainly grains, forest resides, and solid waste. It can be used in its pure form, but is more widely used in a blended form. Gasoline blends using 90% gasoline and 10% ethanol have been widely used in many areas of the country. Ethyl tertiary butyl ether (ETBE) is a feedstock for reformulated gasoline based on ethanol.

By the early 1990s, almost 8% of the gasoline sold in the United States was an ethanol mixture with 850 million gallons of ethanol produced each year. About 95% of this were from the fermentation of corn. Most of this was used as a gasoline additive to produce the 10% ethanol/90% gasoline mixture called gasohol. About 30% of the nation's gasoline had some alcohol in it. Most ethanol use in the United States was in the Midwest, where excess corn and grain crops were used as feedstocks.

In 1979 only 20 million gallons of ethanol were being produced in the United States each year. By 1983, this had jumped to 375 million gallons annually and by 1988 to almost 840 million gallons annually. More than sixty ethanol production facilities were operating by 1993 in the United States in twenty-two states. Farm vehicles were converted to ethanol fuel and demonstration programs were underway for testing light-duty vehicles.

The nation's first E85 (85% ethanol) fueling station opened in La Habra, CA in 1990. The station was operated by the California Renewable Fuels Council.

Although most ethanol is produced from corn, research has been done on producing this type of alcohol fuel from cellulosic biomass including energy crops, forest and agricultural residues, and MSW, which would be much cheaper feedstocks. The process of chemically converting these cellulosic biomass feedstocks is more difficult and until this process can be simplified the price of ethanol remains high.

Ethanol in Brazil

Brazil has been the major producer of ethanol in the world. The Brazilian program to make ethanol from sugarcane began in 1975 and by the 1990s, more than 4 million cars were running on ethanol. The ethanol widely used in Brazil is a mixture of 95% ethanol and 5% water. A small amount (up to 3%) of gasoline is also used.

In Brazil almost 90% of the new cars run on this mixture while the rest operate on a 20% ethanol/80% gasoline mix. The country produces the ethanol on about 1% of its total farmable land. This is because sugarcane can be grown almost year-round in Brazil. The program required government assistance and by 1988 government subsidies for the production of ethanol from sugarcane were almost $1.3 billion.

Brazil's program was not able to supply enough fuel even with more than 15 billion liters produced annually. In order to meet consumer demand, the Brazilian government was forced to import ethanol to meet the demand.

Reformulated Gasoline

Reformulated gasoline is viewed as an alternative fuel that does not require engine modification. Used mainly because its effectiveness in reducing tailpipe emissions, reformulated gasoline was qualified under the Clean Air Act to compete with other alternatives as an option for meeting lower emission standards.

Although the formula can vary by manufacturer, reformulated gasoline usually has polluting components like butane, olefins, and aromatics removed and an octane-enhancer like methyl tertiary butyl ether (MTBE) added. MTBE can reduce carbon monoxide by 9%, hydrocarbons by 4%, and nitrogen oxides by 5%, and improves combustion efficiency. It was widely used in California, Arizona, and Nevada, but is being phased out after it was found to contaminate water supplies.

ARCO, for example, marketed a reformulated gasoline, EC-1 Regular (Emission control-1), for older vehicles without catalytic converters, in southern California. These vehicles made up only a small portion of the car and truck population in the area but they contributed almost a third of the vehicular air pollution. ARCO also marketed a premium reformulated gasoline, EC-Premium. The EPA estimated that the ARCO reformulated gasolines reduced air pollution by about 150 tons each day in southern California.

Natural Gas

Natural gas is a fossil fuel that is found in underground reservoirs. It consists chiefly of methane, with smaller amounts of other hydrocarbons such as ethane, propane, and butane along with inert gases such as carbon dioxide, nitrogen, and helium. The actual composition varies, depending on the region of the source. As an engine fuel, natural gas may be used either in a compressed form, compressed natural gas (CNG) or in a liquid form, liquefied natural gas (LNG).

Although the United States is a major producer and user of natural gas, only a few percent of annual production is used for vehicles, construction and other equipment including power generation. Compressed natural gas is only used in about 30,000 vehicles in the United States, which includes school buses, delivery trucks, and fleet vehicles. Worldwide, about a million vehicles in thirty-five countries operate on natural gas. Some of the countries where natural gas is widely used include New Zealand, Italy and countries of the former Soviet Union.

There are more than 300 NG filling locations in the United States, most of these serve private fleets and about one-third are open to the public. This fuel is more appropriate for fleet vehicles that operate in limited geographical regions and that return to a central location every night for refueling.

In 1991 the California Air Resources Board certified a compressed natural gas (CNG) powered engine as the first alternative fueled engine certified for use in California. The board has also sponsored a test program to fuel school buses with CNG. While CNG has been used for fleet and delivery vehicles, most tanks hold enough fuel for a little over 100 miles.

Natural gas has several advantages over gasoline. It emits at least 40% less hydrocarbons and 30% less carbon dioxide per mile compared to gasoline. It is also less expensive than gasoline on a per gallon-equivalent. Maintenance costs can also be less than those for gasoline engines since natural gas causes less corrosion and engine wear. Although natural gas is a plentiful fossil fuel, it is nonrenewable. There is also a range limitation and natural gas vehicles can cost more due to the need to keep the fuel under pressure. The weight and size the pressure tank reduces storage space and affects fuel economy.

Both methanol and ethanol are alcohol fuels that can be created from renewable sources. Alcohol fuels are converted from biomass or other feedstocks using one or more of the conversion technologies. Government and private research programs are finding more efficient, cost-effective methods of converting biomass to alcohol fuels.

Although methanol was originally a by-product of charcoal production, today it is primarily produced from natural gas, but it can also be made from biomass and coal. When methanol is made from natural gas, the gas reacts with steam to produce synthesis gas, a mixture of hydrogen and carbon monoxide. This then reacts with a catalytic substance at high temperatures and pressures to produce methanol. The process is similar when methanol is produced by the gasification of biomass.

Most of the ethanol in the United States is made from fermenting corn. Dry-milling or wet-milling can be used. In dry-milling, the grain is milled without any separation of its components. The grain is mashed and the starch in the mash is converted to sugar and then to alcohol with yeast.

In wet-milling, the corn is first separated into its major components, the germ, oil, fiber, gluten and starch. The starch is then converted into ethanol. This process provides useful by-products such as corn gluten feed and meal. The only other country with a significant production of ethanol, Brazil, makes its fuel from sugar cane.

One of the arguments regarding the adoption of methanol as a fuel is that it emits higher amounts of formaldehyde, a contributor to ozone formation and a suspected carcinogen, compared to gasoline. Proponents of methanol disagree, saying that only one-third of the formaldehyde from vehicle emissions actually comes from the tailpipe, with the other two-thirds forming photochemically, once the emissions have escaped. They argue that pure methanol vehicles would only generate one tenth as much of the hydrocarbons that are photochemically converted to formaldehyde as do gasoline automobiles.

Methane has a colorless flame and may be explosive in a closed space such as a fuel tank although it is less flammable than gasoline and results in less severe fires when ignited. Colorants can be added to help identify the flame and baffles or flame arresters at the opening of the tank can be used to inhibit the accidental ignition of methanol vapors.

Producing methanol from biomass or coal costs about twice as much as producing it from natural gas. This encourages the use of nonrenewable petrochemical sources over biomass or coal. Considering the full production cycle, methanol from biomass emits less carbon dioxide than ethanol from biomass. This is because short rotation forestry, the feedstocks of methanol, requires the use of less fertilizer and diesel tractor fuel than the agricultural starch and sugar crops which are the feedstocks of ethanol.

The more widespread use of ethanol could have some safety benefits. Ethanol is water soluble, biodegradable, and evaporates easily. Ethanol spills should be much less severe and easier to clean up than petroleum spills.

Agricultural surplus is used for the production of ethanol in the United States and provides economic benefits to farmers and to the farming economy. By 1990, almost 360 million bushels of surplus grain were being used to produce ethanol. It is estimated that, in that year, due to ethanol production, farm income increased by some $750 million, federal farm program costs dropped by $600 million and crude oil imports fell by over forty million barrels.

One of ethanol's major drawbacks in comparison to methanol is its price. It can cost almost twice as much as methanol. But, both methanol and ethanol, as liquids, can use established storage and distribution facilities.

COSTS

Cost differences between gasoline and most alternative fuels present an obstacle to more widespread use of these fuels. While conversion technologies may become more efficient and more cost-competitive over time, as long as gasoline prices remain relatively low, many alternative fuels may not become cost-competitive without government help, in the form of subsidies or tax credits. The cost difference between untaxed renewable fuels and taxed gasoline can be rather small.

By the early 1990s, methanol was about $0.75 per gallon without federal or state tax credits. The cost of wood-derived ethanol dropped from $4.00 to about $1.10 before any tax credits. The federal government provided a tax credit of $0.60 per gallon, which was further subsidized by some states with an additional $0.40 per gallon. These tax credits allowed ethanol to be competitive.

Comparing the per gallon costs of methanol and ethanol with gasoline requires multiplying the gallon cost by the number of gallons needed for the same distance as gasoline. Methanol's energy density is about half that of gasoline, so it takes about two gallons of methanol to get the same amount of power as one gallon of gasoline. A gallon of ethanol contains about two-thirds the energy as a gallon of gasoline.

THE GROWTH OF RENEWABLE FUELS

During the 1920s the catalytic synthesis of methanol was commercialized in Germany. Even before that, methane was distilled from wood. This pyrolysis of wood is a relatively inefficient process. Ethanol saw several periods of popularity in the last century, especially during the world wars when petroleum became limited. In more recent decades, the use of alcohol fuels has seen rapid development.

The worldwide use of MTBE occurred quickly. The first MTBE plant was built in Italy in 1973, and its use then spread through Europe. By 1980, the installed capacity in Europe was almost 90 million gallons per year, which grew to over 300 million gallons per year by the end of 1990. In the United States MTBE production began about 1980 and reached more than a billion gallons by 1987.

Most of the initial interest in alternative fuels started after the oil crisis in the 1970s. It has been grown more recently by concerns about supply interruptions, high prices, air quality and greenhouse gases.

LEGISLATION

In the United States there has been some legislation on developing cleaner-burning gasoline substitutes, gasoline enhancers and more efficient automobiles. The 1988 Alternative Motor Fuels Act (AMFA) and the 1990 amendments to the Clean Air Act (of 1970) were among this legislation. The focus of the AMFA was on demonstration programs that could encourage the use of alternative fuels and alternative-fuel vehicles. The act also offered credits to automakers for producing alternative-fuel vehicles and incentives to encourage federal agencies to use these vehicles.

The 1990 amendments to the Clean Air Act covered a range of issues. New cars sold from 1994 on were to emit about 30% less hydrocarbons and 60% less nitrogen-oxide pollutants from the tailpipe than earlier cars. New cars were also to have diagnostic capabilities for alerting the driver to malfunctioning emission-control equipment. In October 1993 oil refiners were required to reduce the amount of sulfur in diesel fuel. Starting in the winter of 1992/1993, oxygen, to reduce carbon monoxide emissions, was added to all gasoline sold during winter months in any city with carbon monoxide problems.

In 1996 auto companies were to sell 150,000 cars in California that had emission levels of one-half compared with the other new cars. This increased to 300,000 a year in 1999 and in 2001 the emission levels are reduced by half again.

Beginning in 1998 a percentage of new vehicles purchased in centrally fueled fleets in 22 polluted cities were to meet tailpipe standards that were about one-third of those for passenger cars.

TECHNOLOGY ISSUES

If alternative fuels are to be more widely used, changes must take place both in fuel infrastructure, storage and engine technology. Infrastructural changes will improve the availability of alternative fuels. This can be done by modification of existing filling stations and by establishing a distribution system that is as efficient as the current gasoline system.

Technological changes in the manufacture of power sources are required if they are to run on alternative fuels. It is likely that more power sources will move away from single-fuels to several fuels which would compete. This is done in many power plants today. Dual-fuel or flexible-fuel are now used to some degree around the world.

FLEXIBLE FUEL

One of the problems with the development of alternative fuels is the demand question. Why should manufacturers make alternative fuel engines with uncertain fuel supplies? Why should the fuel industry manufacture and distribute fuels without a clear market? Flexible fuel vehicles (FFVs), which are also called variable fuel vehicles, (VFVs) attempt to solve this problem. These vehicles are designed to use several fuels. Most of the major automobile manufacturers have developed FFV prototypes, many of these focus on methanol. These methanol powered vehicles can also use gasoline. There are about 15,000 M85 methanol vehicles in operation. Methanol vehicles can provide greater power and acceleration but they suffer from cold starting difficulties.

Both methanol and ethanol have a lower energy density than that of gasoline and thus more alcohol fuel is needed to provide the same energy. Cold starting problems can occur with these fuels in their pure form, but the addition of a small percentage of gasoline eliminates this problem.

A dual-fuel boiler or engine might operate on natural gas, fuel oil, gasoline or an alternative fuel. Typically, boilers or engines will switch between a liquid or gaseous fuel. Cars, trucks, and buses that use both gasoline and compressed natural gas are being used in northern Italy.

Flexible-fuel engines are able to use a variable mixture of two or more different fuels, as long as they are alike physically, in usually liquid form. Vehicles with flexible-fuel engines are not in widespread use.

Dedicated-fuel vehicles operate on a single fuel which is typically cheaper and more efficient. Vehicles that operate on liquid natural gas are used in taxis in Japan, Korea, and the major difference between compressed natural gas and more conventional fuels is its form. Natural gas is gaseous rather than liquid in its natural state.

Most gasoline-powered engines can be converted to dual-fuel engines with natural gas. The conversion does not require the removal of any of the original equipment. A natural gas pressure tank is added along with a fuel line to the engine through special mixing equipment. A switch selects either gasoline or natural gas/propane operation. Diesel vehicles can also be converted to a dual-fuel configuration.

Natural gas engines may use lean-burn or stoichiometric combustion. Lean-burn combustion is similar to that which occurs in diesel engines, while stoichiometric combustion is more similar to the operation of a gasoline engine.

Compressed natural gas has a high octane rating of 120 and produces 40 to 90% lower hydrocarbon emissions than gasoline. There are also 40 to 90% lower carbon monoxide emissions and 10% lower carbon dioxide emissions than gasoline. A larger, heavier fuel tank is needed and the driver must refill about every 100 miles. Refilling takes two to three times longer than refilling gasoline. Some slow fill stations take several hours and the limited availability of filling stations can be a problem.

FUEL ALCOHOL AND CARBON DIOXIDE

Fuel alcohol programs have been appearing in more and more countries. Energy independence, low market prices for sugar and other food crops, and large agricultural surpluses are the primary reasons for these programs. Several countries with fuel alcohol programs are in Africa and Latin America, along with the United States and a few other countries.

When fuels are derived from biomass, the net increase in carbon dioxide emitted into the atmosphere is usually considered to be neutral or even negative because the plants used to produce the alcohol fuel have reabsorbed the same or more carbon than is emitted from burning the fuel.

The net effect may not be as beneficial when the carbon dioxide emitted by equipment for the harvesting of the biomass feedstocks is considered. This depends on the differences in equipment, farming techniques and other regional factors.

When fuels are produced from biomass, there is job creation in agriculture and related industries. Expanded production can also increase exports of by-products such as corn gluten meal from ethanol.

HYDROGEN FUEL

Hydrogen is abundant, being the most common element in the universe. The sun consumes 600 million tons of it each second. But unlike oil, vast reservoirs of hydrogen are not to be found on earth. The hydrogen atoms are bound together in molecules with other elements and it takes energy to extract the hydrogen so it can be used for combustion or fuel cells. Hydrogen is not a primary energy source, but it can be viewed a means of exchange for getting energy to where it is needed, much like electricity.

Hydrogen is a sustainable, non-polluting source of power that could be used in mobile and stationary applications. As an energy carrier, it could increase our energy diversity and security by reducing our dependence on hydrocarbon-based fuels.

Although hydrogen is the simplest element and most plentiful gas in the universe, it never occurs by itself. It always combines with other elements such as oxygen and carbon. Once it has been separated, hydrogen is an extremely clean-energy carrier. It is clean enough that the U.S. space shuttle program used hydrogen-powered fuel cells to operate the shuttle's electrical systems while one of the by-products was used as drinking water for the crew.

Hydrogen as an alternative to hydrocarbon fuels such as gasoline could have many more potential uses, but it must be relatively safe to manufacture and use. Hydrogen fuel cells can be used to power cars, trucks, electrical plants, and buildings but the absence of an infrastructure for producing, transporting, and storing large quantities of hydrogen could inhibit its growth and practical uses.

Hydrogen can be produced by splitting water (H_2O) into its component parts of hydrogen (H_2) and oxygen (O). One method is the steam reforming of methane from natural gas. Steam reforming converts the methane and other hydrocarbons in natural gas into hydrogen and carbon monoxide using the reaction of steam over a nickel catalyst. Electrolysis uses an electrical current to split water into hydrogen at the cathode (+) and oxygen at the anode (–). Steam electrolysis uses heat, instead of electricity, to provide some of the energy needed to split water and can make the process more energy efficient.

If hydrogen is generated from renewable sources, its production and use can be part of a clean, natural cycle. Thermochemical water splitting uses chemicals and heat in several steps to split water into hydrogen and oxygen. Photolysis is a photoelectrochemical process that uses sunlight and catalysts to split water. Biological and photobiological water splitting use sunlight and biological organisms to split water. Thermal water splitting uses a high temperature of 1000°C to split water. Biomass gasification uses microbes to break down different biomass feedstocks into hydrogen.

Wide-scale Hydrogen Production

Cost is a hurdle in using hydrogen widely as a fuel. Changes in the energy infrastructure are needed to use hydrogen. Electricity is required for many hydrogen production methods. The cost of this electricity tends to make hydrogen more expensive than the fuels it would replace.

Another major concern is hydrogen's flammability. It can ignite in low concentrations and can leak through seals. Leaks in transport and storage equipment could present public safety hazards. These are the practical considerations that need to be addressed before wide-scale use of hydrogen becomes a reality.

Researchers are developing new technologies that can use hydrogen that is stored or produced, as needed, onboard vehicles. These technologies include hydrogen internal combustion engines, which convert hydrogen's chemical energy to electricity using a hydrogen piston engine coupled to a generator in a hybrid electric vehicle.

Onboard reforming for fuel cells, uses catalytic reactions to convert conventional hydrocarbon fuels, such as gasoline or methanol, into hydrogen that fuel cells use to produce electricity to power vehicles.

Hydrogen-based Energy Systems

The announcement of the FreedomCAR Partnership to develop fuel-cell-powered vehicles committed the U.S. Department of Energy toward a hydrogen-based energy system by making fuel-cell-powered vehicles available in 2010. This was a needed push for the development of the technologies needed to make hydrogen-powered transportation a reality.

Fuel Cells

When using hydrogen as fuel, the main emission from fuel cells is potable water. Even when using hydrocarbons as fuel, these systems offer substantial reductions in emissions. Solid Oxide Fuel Cell (SOFC) systems can reach electrical efficiencies over 50% when using natural gas, diesel or biogas. When combined with gas turbines there can be electrical efficiencies of 70%, for small installation as well as large. When using a fuel cell system, these efficiencies can be kept at partial loads as low as 50%, usually conventional technologies must run at nearly full load to be most efficient.

NO_x and SO_x emissions from SOFC systems are negligible. They are typically 0.06 g/kWhe and 0.013 g/kWhe.* SOFCs also produce high-quality heat with their working temperature of 850°C. This makes combined heat and power production possible with SOFC systems. The total efficiency can then reach 85%. Advanced conventional cogeneration of heat and power can reach total efficiencies up to 94% with electrical effi-

---

*Kilo-watt hours electrical

ciencies over 50%. This occurs only at full load. A high electrical efficiency is preferred over heat efficiency, since this results in a higher energy with the initial energy source better utilized, in terms of practical end-use.

Fuel cell systems are modular like computers which makes it possible to build up facilities as needed with parts in an idle mode when full capacity is not needed. The capacity is easily adjusted, as the need arise.

Hydrocarbons such as natural gas or methane can be reformed internally in the SOFC, which means that these fuels can be fed to the cells directly. Other types of fuel cells require external reforming. The reforming equipment is size-dependent which reduces the modularity.

Hydrogen Fuel Cells and Global Warming

Unlike internal combustion engines, hydrogen fuel cells do not emit carbon dioxide. But, extracting hydrogen from natural gas, gasoline or other products requires energy and involves other by-products.

Obtaining hydrogen from water through electrolysis consumes large amounts of electrical power. If that power comes from plants burning fossil fuels, the end product can be clean hydrogen, but the process used to obtain it can be polluting. After the hydrogen is extracted, it must be compressed and transported, if the machinery and vehicles run on fossil fuels, they will produce CO_2. Running an engine with hydrogen extracted from natural gas or water could produce a net increase of CO_2 in the atmosphere.

Fuel cell cars must be able to drive hundreds of miles on a single tank of hydrogen. A gallon of gasoline contains about 2,600 times the energy of a gallon of hydrogen. If hydrogen cars are to travel 300 miles on a single tank, they will have to use compressed hydrogen gas at very high pressures, up to 10,000 pounds per square inch. Even at this pressure, cars would need large fuel tanks.

Liquid hydrogen may be better. The GM liquid-fueled HydroGen3 goes 250 miles on a tank about twice of size of a typical gasoline tank. The car must be driven every day to keep the liquid hydrogen chilled to –253°C or it boils off.

The future prospects for a hydrogen economy were explored in "The Hydro Economy: Opportunities, Costs, Barriers, and R&D Needs." This government-sponsored study was published in 2004 by the National Research Council. ExxonMobile, Ford, DuPont, the Natural Resources Defense Council and others contributed to the report. It urged more stringent tailpipe-emission standards and more R&D funding for renewable energy and alternative fuels. It also recommended that the Department of Energy balance a portfolio of R&D efforts and explore alternatives.

The hydrogen economy could arrive by the end of the next decade or closer to mid-century. But, interim technologies will play a critical role in the transition. One of the most important of these technologies is the gas-electric hybrid vehicle, which uses both an internal combustion engine and an electric motor. Electronic power controls allow switching almost seamlessly between these two power sources to optimize gas mileage and engine efficiency. U.S. sales of hybrid cars has been growing steadily, and the 2005 models included the first hybrid SUVs, Ford Escape, Toyota Highlander and Lexus RX400h.

Researchers sponsored by the FreedomCAR program are investigating ultralight materials, which include plastics, fiberglass, titanium, magnesium, carbon fiber and developing lighter engines made from aluminum and ceramic materials. These new materials can reduce power requirements and allow other fuels and fuel cells to become popular more quickly.

The additional costs to manufacture vehicles that run on alternative fuels has been a subject of much debate. Many believe that when all changes have been taken into account, the costs for near alcohol automobiles will be very close to the cost of a gasoline automobile. FFVs are expected to cost slightly more.

EPA estimates show that, with the necessary adjustments, the savings and costs will balance out to zero. The increased costs necessary for fuel tank adjustments and to compensate for cold-start problems could be balanced out by smaller, lighter engines that use near fuel these cars can have because of their increased efficiency.

The case is different with dual fuel engines that could use compressed natural gas and a liquid fuel. Engines can be converted to run on compressed natural gas at a cost of several thousand dollars.

A portfolio of energy-efficient technologies can help to release us from fossil fuels. If consumers had a wider and more diverse set of energy sources, the economy could be more robust and the world could be more stable.

Applications for Hydrogen Fuel Cells

Cars and light trucks produce about 20% of the carbon dioxide emitted in the U.S., while power plants burning fossil fuels are responsible for more than 40% of CO_2 emissions. Fuel cells can be used to generate electricity for homes and businesses. Plug Power, UTC, FuelCell Energy and Ballard Power Systems already produce stationary fuel cell generators. Plug Power has hundreds of systems in the U.S. including the first fuel-cell-powered McDonald's. The installed fuel cells have a peak generating capacity of less than 100 megawatts, which is 0.01% of the almost one million megawatts of U.S. generating capacity.

Hydrogen R&D

President George W. Bush pledged to spend $1.2 billion on hydrogen yet the Department of Energy spends more on nuclear and fossil fuel research than on hydrogen. The government's FreedomCAR program, funds hydrogen R&D in conjunction with American car manufacturers. The program requires that the companies demonstrate a hydrogen-powered car by 2008.

The Center for Energy, Environmental and Economic Systems Analysis at Argonne National Laboratory near Chicago estimates that building a hydrogen economy would take more than $500 billion.

Oil companies are not willing to invest in production and distribution facilities for hydrogen fueling until there are enough hydrogen cars on the road. Automakers will not produce large numbers of hydrogen cars until drivers have somewhere to fill them up.

The Department of Energy's hydrogen-production research groups reports that a fourth to a third of all filling stations in the U.S. would be needed to offer hydrogen before fuel cells become viable as vehicle power. California has its Hydrogen Highway Project with 150 to 200 stations at a cost of about $500,000 each. These would be situated along the state's major highways by 2010. There are over 100,000 filling stations in the U.S. Retrofitting just 25% of those with hydrogen fueling systems could cost more than $12.5 billion.

Iceland and Hydrogen

Iceland's first hydrogen fueling station is operating near Reykjavik. The hydrogen powers a small fleet of fuel cell buses and is produced on-site from electrolyzed tap water. The Iceland New Energy consortium includes automakers, Royal Dutch/Shell and the Icelandic power company Norak Hydro. It plans to convert the rest of the nation to hydrogen.

Almost 75% of Iceland's electricity comes from geothermal and hydroelectric power. This available clean energy allows Iceland to electrolyze water with electricity from the national power grid. In the U.S. only about 15% of grid electricity comes from geothermal and hydroelectric sources, while 71% is generated from fossil fuels.

Only 16 hydrogen fueling stations are planned to allow Icelanders to refuel fuel cell cars around the country. At almost 90 times the size of Iceland, the U.S. could start with about 1,500 fueling stations. This assumes that the stations are placed properly to cover the entire U.S. with no overlap. The stations would cost about $7.5B.

Hydrogen Cars

Volume production of fuel cell cars should reduce costs, but one Department of Energy projection with a production of 500,000 vehicles a year still has the cost too high. However, efforts continue to improve fuel cell technology and utilization which should reduce costs. The General Motors fuel cell program aims at having a commercial fuel cell vehicle by 2010.

A California company called HaveBlue sells a power system for sailing yachts with solar panels, a wind generator and a fuel cell. The solar panels provide 400 watts of power for the cabin systems and electrolyzer for producing hydrogen from salt or fresh water. The hydrogen is stored in six tanks in the keel. Up to 17 kilograms of hydrogen is stored in solid matrix metal hydrid. The tanks replace 3,000 pounds of lead ballast.

The wind generator has an output of 90 watts under peak winds and starts producing power at 5 knots of wind. The fuel cell produces 10 kilowatts of electricity along with steam which is used to raise the temperature of the hydrogen storage tanks. A reverse-osmosis water system desalinates water for cabin use and a deionizing filter makes pure water for fuel cell use.

A potential problem with the proton exchange membrane (PEM) fuel cell, which is the type being developed for automobiles is life span. Internal combustion engines have an average life span of 15 years, or about 170,000 miles. Membrane deterioration can cause PEM fuel cells to fail after 2,000 hours or less than 100,000 miles.

Ballard's original PEM design has been the prototype for most automobile development. This has been the basic design that is used to demonstrate fuel cell power in automobiles. But, it may not be the best architecture and geometry for commercial automobiles. The present geometry may be keeping the price up. Commercial applications require a design that will allow economies of scale to push the price down.

Hydrogen Production

Nuclear power can produce hydrogen without emitting carbon dioxide into the atmosphere. Electricity from a nuclear plant would electrolyze water splitting H_2O into hydrogen and oxygen. However that nuclear power can create long-term waste problems and has not been economical. One study done by the Massachusetts Institute of Technology and Harvard University, concluded that hydrogen produced by electrolysis of water will depend on low cost nuclear power.

Performing electrolysis with renewable energy, such as solar or wind power eliminates the pollution problems of fossil fuels and nuclear power. However, current renewable sources only provide a small fraction of the energy that is needed for a hydrogen fuel supply.

From 1998 to 2003, the generating capacity of wind power increased 28% in the U.S. to about 6,500 megawatts, enough for less than 2 million homes. Wind is expected to provide about 6% of the nation's power by 2020.

The University of Warwick in England estimated that converting every vehicle in the U.S. to hydrogen power would require the electricity output of a million wind turbines enough to cover half of California. Solar panels would also require huge areas of land.

Water may be another factor for hydrogen production, especially in the sunny regions most suitable for solar power. A study by the World Resources Institute in Washington, D.C. estimates that obtaining enough hydrogen with electrolysis would require over 4 trillion gallons of water yearly. This is about the flow over Niagara Falls every 90 days. Water consumption in the U.S. could increase by about 10%.

Hydrogen Leakage

Hydrogen gas is odorless and colorless, and it burns almost invisibly. A fire may not be detected until it is too late. It does not take much to set off compressed hydrogen gas. A cell phone may provide enough of a static discharge to ignite hydrogen.

An accident may not cause an explosion, since carbon fiber reinforced hydrogen tanks are almost indestructible. But, there is always the danger of leaks in fuel cells, refineries, pipelines and fueling stations. Hydrogen is a gas, while most of our other fuels are liquids.

A high-pressure gas or cryogenic liquid hydrogen fuel distribution would be much different. Hydrogen is such a small molecule that it tends to leak through the finest of openings.

A leaky infrastructure could affect the atmosphere. Researchers from the California Institute of Technology and the Jet Propulsion Laboratory in Pasadena, CA, compared statistics for accidental industrial hydrogen and natural gas leakage. These were estimated at 10 to 20% of total volume. Extending these estimates to an economy that runs on hydrogen could result in four to eight times as much hydrogen in the atmosphere.

The Department of Energy's Office of Energy Efficiency and Renewable Energy thinks these estimates are much too high, but whatever the volume, more hydrogen in the atmosphere will then combine with oxygen to form water vapor, creating more clouds. The increased cloud cover could affect the weather and global warming.

Giant molecular cloud formations, consisting almost entirely of hydrogen, are the most massive objects within galaxies. Gravity eventually causes the hydrogen to compress until it fuses into heavier elements.

Without the energy emitted by the sun, life as we know it could not exist. We know that the primary fuel for the sun and other stars is hydrogen. Although the force that causes the sun and other stars to burn is gravity, the fuel is hydrogen.

Our sun consumes about 600 million tons of hydrogen every second. As this hydrogen is fused into helium, photons of electromagnetic energy are released and eventually find there way through the earth's atmosphere as solar energy. This solar energy is the aftermath of nuclear fusion, while nuclear fission occurs in commercial nuclear reactors. Without this energy there would be no life, there would be no fossil fuels or wind or elements like uranium.

If there is a need to make a transition from nonrenewable fossil fuel, then we should consider the development of technologies that can use the available energy of the sun. It is rational to suppose that solar energy will eventually serve as a primary energy source.

Protons and electrons are the basic components of the hydrogen atom and these atoms are the basic building blocks of the other 91 elements that occur naturally. The atomic number of an atom equals the number of protons, hydrogen nuclei, or electrons of the element.

Since hydrogen has one proton and one electron, it has an atomic number of 1. Carbon has six protons and six electrons and an atomic number of 6. The proton's positive electrical charge and the electron's negative charge have a natural attraction for each other.

The elements were probably formed during the first few seconds of the origin of the universe following the big bang theory which took place some 15 billion years ago. Gravity is the basic universal force that causes every particle of matter to be attracted to every other particle. Hydrogen atoms and other subatomic particles would have continued to expand away from each other from the force of the big bang, but gravity caused these particles to cluster in large masses. As the mass increased, the force of gravity increased.

Eventually, the force and pressure became great enough for the interstellar clouds of hydrogen to collapse causing the hydrogen and other particles to collide. The collisions result in high enough temperatures of 45 million degrees Fahrenheit and pressures to fuse the hydrogen into helium which is the birth of a star. As a star feeds on this supply of hydrogen, four hydrogen nuclei are fused into one heavier helium nucleus.

The heavier helium atoms form a dense, hot core. When the star has consumed most of its hydrogen, it begins to burn or fuse the helium, converting it to carbon and then to oxygen.

The more massive a star is, the higher the central temperatures and pressures are in the later stages. When the helium is consumed, the star fuses the carbon and oxygen into heavier atoms of neon, magnesium, silicon and even silver and gold. In this way, all the elements of the earth except hydrogen and some helium were formed billions of years ago in stars.

As we attempt to use solar energy to replace the use of fossil and nuclear fuels, this relationship between solar energy and hydrogen returns and one may not effectively work without the other. Hydrogen utilization requires some type of a primary energy input to separate it from the other atoms of oxygen or carbon. Solar energy will not be able to replace fossil or nuclear-fueled energy systems unless it can be efficiently stored, transported and used as a combustion fuel in vehicles and power plants.

Water Former

Hydrogen was discovered in 1766 when the English chemist Henry Cavendish observed what he called an inflammable air rising from a zinc-sulfuric acid mixture. It was identified and named in the 18th century by Antoine Lavoisier, who demonstrated that this inflammable air would burn in air to form water. He identified it as a true element, and called it hydrogen, which is Greek for water former.

Hydrogen is the simplest, lightest and most abundant of the 92 elements in the universe. Making up over 90% of the universe and 60% of the human body. As the most basic element, it can never be exhausted since it recycles in a relatively short time. If hydrogen was made readily available for electric power generation instead of fossil fuels, electricity costs could be reduced.

Hydrogen can be burned in a combustion chamber instead of a conventional boiler, so high-pressure superheated steam can be generated and fed directly into a turbine. This could cut the capital cost of a power plant by one half.

When hydrogen is burned, essentially no pollution is generated. Expensive pollution control systems, which can be almost one third of the capital costs of conventional fossil fuel power plants are not required. This should also allow plants to be located closer to residential and commercial loads, reducing power transmission costs and line losses.

Since hydrogen burns cleanly and reacts completely with oxygen to produce water vapor, this makes a more desirable fuel than fossil fuels for essentially all industrial processes. For example, the direct reduction of iron or copper ores could be done with hydrogen rather than smelting by coal or oil in a blast furnace. Hydrogen can be used with conventional vented burners as well as unvented burners. This would allow utilization of almost all of the 30 to 40% of the combustion energy of conventional burners that is lost as vented heat and combustion by-products.

Universal Fuel

Hydrogen is different than other energy options like oil, coal, nuclear or solar. Solar technology is renewable, modular and generally pollution free, but it has some disadvantages, such as not always being available at the right time.

Hydrogen is a primary chemical feedstock in the production of fuels including gasoline, lubricants, fertilizers, plastics, paints, detergents, electronics and pharmaceutical products. It is also an excellent metallurgical refining agent and an important food preservative.

Hydrogen can be extracted from a range of sources since it is in almost everything, from biological tissue and DNA, to petroleum, gasoline, paper, human waste and water. It can be generated from nuclear plants, solar plants, wind plants, ocean thermal power plants or green plants.

Hydrogen and electricity are complementary and one can be converted into the other. Hydrogen can be viewed as a type of energy currency that does not vary in quality depending on origin or location. A molecule of hydrogen made by the electrolysis of water is the same as hydrogen manufactured from green plant biomass, paper, coal gasification or natural gas.

Primary and Secondary Energy Sources

Hydrogen is often called a secondary energy carrier, instead of a primary energy source. This is because energy must be used to extract the hydrogen from water, natural gas, or other compound that contains the hydrogen. This classification is not exact since it assumes solar, coal, oil or nuclear are primary energy sources, meaning that energy is not needed to obtain them.

However, finding, extracting and delivering these so-called primary energy sources requires energy and major investments before they can be utilized. Coal and natural gas come closer to true primary energy sources since they can be burned directly with little or no refining, but energy is still needed to extract these resources and deliver them where the energy is needed. Even when extensive drilling for oil is not required from shallow wells or pools, energy must still be used for pumping and refining.

Many environmental problems can result from finding, transporting and burning fossil fuels. But, when hydrogen is used as a fuel, its by-product is essentially water vapor. When hydrogen is burned in air, which contains nitrogen, nitrogen oxides can be formed as they are in gasoline engines. These oxides can almost be eliminated in hydrogen engines by lowering the combustion temperature of the engine.

Some tests have shown that the air coming out of a hydrogen fueled engine is cleaner than the air entering the engine. Acid rain, ozone depletion and carbon dioxide accumulations could be greatly reduced by the use of hydrogen.

Hydrogen Storage

Hydrogen can be stored as a gas, liquid, or as a part of a solid metal, polymer or liquid hydrid. Studies have indicated that large-scale storage could take place with gaseous hydrogen underground in aquifers, depleted petroleum or natural gas reservoirs or man made caverns from mining operations.

One of the obstacles in using hydrogen as an automotive fuel is storing it safely and efficiently on board vehicles. Although it is possible to store hydrogen as a high pressure gas in steel containers, disadvantages exist because of the weight of the storage containers and the safety hazard in the event of an accident. Other methods of storage for hydrogen include solid or liquid hydrids, low temperature cryogenic liquids, or a combination of the two.

Hydrogen Hydrids

Hydrid materials will absorb hydrogen like a sponge, and then release it when heated. There are hundreds of hydrid materials. The first hydrid systems used in automotive vehicles consisted of metal particles of iron and titanium that were developed at Brookhaven National Laboratory. These were tested by Daimler-Benz in Stuttgart, Germany. These early hydrid systems were shown to be safe for storing hydrogen in automobiles, but they are almost 5 times heavier than liquid hydrogen storage systems.

Other hydrid systems do not have such weight penalties and include magnesium nickel alloys, non-metallic polymers, or liquid hydrid systems that use engine heat to disassociate fuels like methanol into a mixture of hydrogen and carbon monoxide.

In an iron titanium hydrid system, for a range of 300 miles (480 kilometers), the tank could weigh about 5,600 pounds (2,520 kilograms). A liquid hydrogen tank for this range would weigh about 300 pounds (136 kilograms), a comparable gasoline tank would weigh about 140 pounds (63 kilograms).

An electric vehicle with a similar range and lead acid batteries would have a battery weight of about 6,500 pounds (2,925 kilograms). More efficient battery systems are becoming available but the most efficient electric vehicles of the future may be energized by fuel cell systems that convert hydrogen and oxygen directly into electricity. These systems would depend on having hydrogen fuel more readily available.

Liquid Hydrogen

When hydrogen gas is liquefied, it needs to be cooled to −421.6° Fahrenheit, making liquid hydrogen a cryogenic fuel. Cryogenics is the study of low temperature physics. A beaker of liquid hydrogen at room temperature will boil as if it was on a hot stove. If the beaker of liquid hydrogen is spilled on the floor, it is vaporized and dissipates in a few seconds. If liquid hydrogen is poured on the hand, it would feel cool to the touch as it slides through the fingers. This is due to the thermal barrier that is provided by the skin. But, place a finger in a vessel containing liquid hydrogen and severe injury will occur in seconds because of the extremely cold temperature. This hydrogen fuel on board a vehicle would allow the use of a small, efficient fuel cell Stirling engine cryocooler system to provide air conditioning.

In most accidents, the most serious concern would be a fuel fed fire or explosion. In this case, liquid hydrogen is generally considered to be a preferred fuel.

Liquid hydrogen is a fuel option that could be utilized on a large scale since it most resembles gasoline in terms of space and weight. Although a liquid hydrogen storage tank for a vehicle could weigh about five times heavier in dry weight than a 30 pound gasoline tank, in vehicles that carry greater volumes of fuel, such as trucks or trains or aircraft, the difference in tank weight could be more than offset by the difference in fuel weight. Studies by Lockheed Aircraft have shown that a large commercial aircraft could have its overall takeoff weight reduced by as much as 40% if liquid hydrogen were used instead of aviation fuel. Liquid hydrogen has the lowest weight per unit of energy, with relatively simple supply logistics with normal refuel times and is generally safer than gasoline in accidents.

Cryogenic fuels like liquid hydrogen are more difficult to handle and substantially more difficult to store compared to hydrocarbon fuels like gasoline or aviation kerosene. Even with highly-insulated double-walled, vacuum-jacketed storage tanks liquid hydrogen can evaporate at a rate of almost 9% per day.

This evaporation increases the pressure on the tank wall and the gaseous hydrogen must be vented to the atmosphere to keep the tank from rupturing. During tests at the Los Alamos National Laboratory a liquid hydrogen fueled vehicle tank of liquid hydrogen evaporated away in about 10 days.

This venting of the fuel must be done to keep a fuel tank full when refueling. In an enclosed space, the vented hydrogen also presents a risk because of hydrogen's wide flammability limits. Hydrogen explosions are rare, but any combustible gas in an enclosed space can be a safety problem. One solution is to burn off the escaping hydrogen and use this energy for heating or cooling. It can also be used to power a fuel cell.

Stationary liquid hydrogen storage tanks that are used in laboratories are able to keep the hydrogen in a liquid state for several months. It should be possible to build vehicular storage tanks that would maintain hydrogen in a liquid state for several weeks. The small quantity of hydrogen evaporating from such tanks could also be sent to a fuel cell that would use the hydrogen to generate electricity. It is also possible to vent the vaporized hydrogen gas to an auxiliary hydrid system for storage.

The double walled vacuum jacketed storage tanks and piping that are required for liquid hydrogen are expensive compared to conventional fuel storage tanks. A gasoline tank might cost about $150, while a liquid hydrogen storage tank could cost a few thousand dollars. Because of the energy density of liquid hydrogen, it requires a fuel tank that is three to four times as large in volume as required for gasoline or aviation fuel.

Liquid hydrogen fuel systems would require changes in the energy infrastructure and end use systems, such as stoves, engines and fueling systems. While disadvantages of liquid hydrogen are substantial, they can be minimized. A few thousand dollars for a liquid hydrogen storage tank seems high, but consider that the emissions control equipment required on gasoline fueled engines adds much to the cost of current vehicles. As production volumes of cryogenic storage tanks increase, the cost of cryogenic tanks are expected to drop below $1,000.

Although cryogenic fuels are difficult to handle, a self-service liquid hydrogen pumping station was built decades ago at Los Alamos National Laboratory. It was shown to be feasible for refueling vehicles over an extended period of time without any major difficulties.

While the increased costs associated with making a changeover to hydrogen energy systems seem high, remember that the environmental costs of finding, transporting and burning fossil fuels are not calculated in the current energy pricing structure. The costs of atmospheric pollution are billions of dollars in additional health care costs, forests and crop losses and the corrosion of buildings and other structures.

Hydrogen Engines

Many engineering groups in the U.S., Germany, Japan, France and other countries are involved in hydrogen research and development. Hydrogen fueled engines tend to be more energy efficient because of their complete combustion. Gasoline and diesel engines form carbon deposits and acids that erode the interior surfaces of the engine and contaminate the engine oil. This increases wear and corrosion of the bearing surfaces. Since hydrogen engines produce no carbon deposits or acids, they should require far less maintenance. Hydrogen can also be used in more efficient Stirling cycle engines.

In the 1920s, a German engineer, Rudolf Erren, began optimizing internal combustion engines to use hydrogen. Erren modified many trucks and buses. A captured German submarine in World War II had a hydrogen engine and hydrogen powered torpedoes that were designed and patented by Erren.

The first hydrogen automobile in the U.S. was a Model A Ford truck, modified in 1966 by Roger Billings while he was a student in high school.

A few years later as a student at Brigham Young University, he won a 1972 Urban Vehicle Design Contest with a hydrogen Volkswagen. Billings established Billings Energy Corporation in Provo, Utah and went on to modify a wide range of vehicles, including a Winnebago motor home that had the engine fueled by hydrogen as well as the generator and appliances. Billings has also constructed a hydrogen home which had the appliances modified to operate on hydrogen.

Most of these vehicles are dual fueled and run on hydrogen or gasoline. The driver is able to change from hydrogen to gasoline while driving with a switch from the vehicle.

Special burner heads have been used by the Tappan Company for hydrogen combustion in stoves. Since hydrogen burns with an invisible flame, Tappan used a steel wool catalyst that rests on the burner head. The stainless steel mesh glows when heated and resembles an electric range surface when the burner is on. Billings also adapted a Coleman Stove for hydrogen. A small hydrogen storage tank with iron-titanium metal hydrids was used.

Hydrogen research programs were also initiated in the U.S. Air Force, Navy and the Army in the 1940s when fuel supplies were a concern. After World War II and prior to the Arab oil embargo in 1973, oil was selling for less than $3 per barrel. The fuel supply problem was not a concern. During the Arab oil embargo of 1973, there were long gas lines in the U.S. and the price of oil quadrupled. This started renewed research into alternative energy supplies including solar power.

Hydrogen Safety

Many believe that hydrogen is particularly dangerous. There are some that think hydrogen energy is related to the hydrogen bomb. But, hydrogen used as a fuel involves a simple chemical reaction involving the transfer of electrons to produce an electric current while a hydrogen bomb requires a high temperature nuclear fusion reaction similar to that which occurs in our sun and other stars.

Others recall that the German airship the Hindenburg used hydrogen when it burst into fire in 1937. While 35 people lost their lives another 62 others survived. Before its crash in 1937, the Hindenburg had successfully completed 10 round trips between the U.S. and Europe. Its sister ship, the Graf Zeppelin, made regular scheduled transatlantic crossings from 1928 to 1939 with no accidents. There were 161 rigid airships that flew between 1897 and 1940, almost all of these used hydrogen. Only 20 were destroyed by fires. Of these 20, seventeen were lost in military action that in many cases the fires resulted from enemy fire during World War I.

Hydrogen has a wider range of flammability when compared to gasoline. A mixture as low as 4% hydrogen in air, or as high as 74% will burn, while the fuel to air ratios for gasoline only range from 1 to 7.6%. It also takes very little energy to ignite a hydrogen flame, about 20 micro-joules, compared to gasoline which requires 240 micro-joules. However, these hazardous characteristics are reduced by the fact that as the lightest of all elements, hydrogen has a very small specific gravity.

Diffusion Rate

Since the diffusion rate of a gas is inversely proportional to the square root of its specific gravity, the period of time in which hydrogen and oxygen are in a combustible mixture is much shorter than other hydrocarbon fuels. The lighter the element is, the more rapidly it disperses when it is released in the atmosphere.

In a crash or accident where hydrogen is released, it rapidly disperses up and away from the ground and any combustible material within the area. Gasoline and other hydrocarbon fuels are heavier since the hydrogen is bonded to carbon which is a much heavier element.

When hydrocarbon fuels vaporize, their gases tend to sink rather than rise into the atmosphere. This allows burning gasoline to cover objects and burn them. In most accidents, hydrogen would be a more desirable fuel.

On March 27, 1977, two fully-loaded Boeing 747 commercial aircraft crashed into each other on a foggy runway in the Canary Islands. This disaster was then the worst in aviation history and took 583 lives. An investigation concluded most of the deaths in the Canary Islands accident resulted from the aviation fuel fire that lasted for more than 10 hours. G. Daniel Brewer, who was the hydrogen program manager for Lockheed, stated that if both aircraft had been using liquid hydrogen fuel instead of kerosene, hundreds of lives would have been saved. He listed a number of reasons.

The liquid hydrogen would not react with oxygen and burn until it first vaporized into a gas. As it evaporated, it would have dissipated rapidly as it was released in the open air. This means that the fuel fed portion of the fire would have only lasted for several minutes instead of hours.

The hydrogen fire would have been confined to a relatively small area because the liquid hydrogen would rapidly vaporize and disperse into the air, burning upward, instead of spreading like aviation fuel.

The heat radiated from the hydrogen fire would be considerably less than that generated by a hydrocarbon fire and only objects immediately adjacent to the flames would be affected. A hydrogen fire produces no smoke or toxic fumes, which in many cases is the cause of death in fires.

In liquid hydrogen fuel storage tanks, the gaseous hydrogen that vaporizes fills the empty volume inside the tanks. This hydrogen is not combustible since no oxygen is present. In gasoline or other hydrocarbon fuel tanks, air fills the empty volume of the tanks and combines with vapors from the fuel to create a combustible mixture.

On September 11, 2001, two fully-loaded Boeing 747 commercial aircraft were hijacked and flown into the World Trade Center. Over 3,000 were killed as the fires inside the twin towers caused the building to collapse. If hydrogen was used as the fuel the damage would have been confined to the immediate crash sites, the buildings would probably be still standing and many lives would have been spared.

A hydrogen fueled vehicle could be fueled by vacuum jacketed liquid hydrogen storage tanks. Vacuum jacketed cryogenic fuel lines carry the liquid hydrogen from the storage tanks. One of the two lines, taps off the gaseous hydrogen displaced from the fuel tank by the incoming liquid hydrogen for returning to the liquefaction plant.

The studies by Lockheed found that along with hydrogen's safety characteristics, liquid hydrogen fueled aircraft would be lighter, quieter, with smaller wing areas and could use shorter runways. Pollution would be much less and the range of an aircraft could be almost doubled, even though the takeoff weight remain about the same.

Hydrogen Explosions

The Hindenburg did not explode, it caught fire. The flames spread rapidly and the airship sank to the ground. The fire was started when the airship was venting some of its hydrogen, to get closer to the ground, during an electrical thunderstorm. The airship was also moored to the ground by a steel cable, which acts as an antenna for electrical discharges.

Hydrogen explosions can be powerful when they occur, but they are rare. Hydrogen must be in a confined space for an explosion to occur. In the open it is difficult to cause a hydrogen explosion without using heavy blasting caps.

In 1974, NASA examined 96 accidents or incidents involving hydrogen. At this time, NASA tanker trailers had moved more than 16 million gallons of liquid hydrogen for the Apollo-Saturn program. There were five highway accidents that involved extensive damage to the liquid hydrogen transport vehicles. If gasoline or aviation fuel had been used, a spectacular fire would have resulted, but none of these accidents caused a hydrogen explosion or fire.

At Wright-Patterson Air Force Base, armor-piercing incendiary and fragment simulator bullets were fired into aluminum storage tanks containing both kerosene and liquid hydrogen. The test results indicated that the liquid hydrogen was safer than conventional aviation kerosene.

Other tests have involved simulated lightning strikes, with a 6-million volt generator that fired electrical arcs into the liquid hydrogen containers. None of these tests caused the liquid hydrogen to explode. Fires did occur from the simulated lightning strikes, but the fires were less severe even though the total heat content of the hydrogen was twice that of kerosene. These tests indicated that liquid hydrogen would be safer than fossil fuels in combat where a fuel tank could be penetrated.

A well publicized event where explosive mixtures of hydrogen and oxygen were present in a confined space occurred during the events in 1979 at the Three Mile Island (TMI) nuclear facility in Pennsylvania. Nuclear reactors operate at very high temperatures. To prevent their six to eight inch thick steel reactor vessels from melting, large amounts of cooling water are continuously circulated in and around the reactor vessel.

An average commercial-sized reactor requires about 350,000 gallons of water per minute. During the process of nuclear fission, the center of the uranium fuel pellets in the fuel rods can reach 5,000°F. The cooling water keeps the surface temperature of the pellets down to about 600°. If the circulating water is not present, in 30 seconds the temperatures in the reactor vessel can be over 5,000°. This temperature is high enough to melt steel and thermochemically split any water present into an explosive mixture of hydrogen and oxygen. This is what happened at TMI. If a spark had ignited the hydrogen gas bubble that drifted to the top of the containment building, the resulting explosion could have fractured the walls. This would have resulted in the release of large amounts of radiation at ground level.

The hydrogen gas bubble was vented, since as long as it remained in the confined space of the containment building, the potential for detonation existed.

A hydrogen gas bubble developing from a nuclear reactor accident is a highly unusual event and is an example of the particular environment that is required for hydrogen to explode.

NASA and Hydrogen

Dr. Warner Von Braun was a German rocket engineer who helped to develop the V-2 rockets in World War II. He was involved in the first efforts to use liquid hydrogen as a rocket fuel. After the war, Von Braun had a major part in the U.S. space program, which evolved into the National Aeronautics and Space Administration (NASA). Since liquid hydrogen has the greatest energy content per unit weight of any fuel, NASA engineers used liquid hydrogen as the primary fuel for the Saturn 5 moon rockets and the Space Shuttle.

NASA also funded research by several aerospace firms, including Lockheed and Boeing, to determine if liquid hydrogen could be practical for commercial aircraft and what modifications would be needed for airports and fueling systems.

The Space Shuttle's main liquid hydrogen-oxygen tank is the largest of the three external tanks. The two smaller boosters use a solid aluminum based fuel.

NASA has been using large quantities of gaseous and liquid hydrogen for many years, which required developing the necessary pipelines, storage tanks, barges and transport vehicles. As a result of this experience, NASA has concluded that hydrogen can be as safe or in some ways safer, than gasoline or conventional aviation fuels.

NASA engineers originally wanted to develop a reusable manned liquid hydrogen-fueled launch vehicle for the space shuttle program, but Congress would not vote for the additional funds that would be needed. Less expensive solid rocket boosters were used, which turned into a tragedy when one of the seals of the solid rocket boosters failed during a cold weather launch. This caused the explosion of the Challenger shuttle in 1986 and the loss of its entire crew, including the first teacher on a spaceflight.

Today's Hydrogen

Most of the hydrogen that is manufactured for industry is made by reacting natural gas with high temperature steam, to separate the hydrogen from the carbon. But, manufacturing hydrogen from fossil fuel resources does not solve the fossil fuel depletion problem.

Making hydrogen from water through electrolysis was initially promoted by nuclear engineers who thought that nuclear generated power would be inexpensive enough to make hydrogen.

References

Behar, Michael, "Warning: the Hydrogen Economy May Be More Distant Than It Appears," *Popular Science*, Volume 266 Number 1, January 2005, pp. 65-68.

Braun, Harry, *The Phoenix Project: An Energy Transition to Renewable Resources*, Research Analysts: Phoenix, AZ, 1990.

Carless, Jennifer, *Renewable Energy*, Walker and Company: New York, 1993.

Cothran, Helen, Book Editor, *Global Resources: Opposing Viewpoints*, Greenhaven Press,: San Diego, CA, 2003.

Romm, Joseph J., *The Hype About Hydrogen*, Island Press: Washington, Covelo, London, 2004.

CHAPTER 2

THE EVOLUTION OF OIL

In 1808 a scientific expedition from the Imperial Academy of Sciences of St. Petersburg declared that petroleum is a mineral of no usefulness. Today, about 2 trillion barrels are utilized annually for a number of purposes.

Petroleum and its derivatives have been put to various uses for a long time. The soil of the Middle East has always been known to be impregnated with oil, and in Chaldea, Egypt and China the distillation of petroleum was familiar several thousands of years ago. It was used in lighting, in the treatment of many ailments, and for purposes of war. Marco Polo writes of petroleum in his voyage across Asia.

The word petroleum did not exist until the Renaissance. Rock-oil, mineral oil, or naphtha, were the names given to the mineral before that time. Petroleum appears in a sixteenth, century book, as an oil which oozes out of rocks on the estate of the Duke of Ferrara, near Modena. An engraving with the text shows a gushing spring and people filling vessels from it. This book also explains some of the uses of petroleum. The oil was used to cleanse an ulceration and heal old wounds. It was also used as balm for burns and bruises.

The shell of a filbert or hazelnut was filled with a little petroleum and downed with a goblet of warm beer for colds or internal problems. Dog bites and stings by a poisonous animal were treated by rubbing the wound or bite with petroleum.

Petroleum was used for many ills including coughs, bronchitis, pulmonary congestion, cramp, gout, rheumatism and eye strain. Modena petroleum was sold as far away as France, where it competed with local oils.

Rock oil, mineral oil, or the liquid ore as it was called, had a well-established reputation, as a therapeutic product in the Middle Ages and in the succeeding centuries. Only recently did it come to be utilized as a motor fuel. After its origin in early European and Asian annals it became an essentially American product before assuming its current world importance.

In the 18th century, the story of American petroleum begins with the first contacts between settlers and Indians. Black oil collected from the surface of the marshes was traded for glass beads and alcohol. Colonists used the petroleum to grease the axles of wagons, heal the wounds of horses, and to treat rheumatism and injuries.

About 1800 drysalters began selling the elixir of life. Most shallow wells in Pennsylvania would fill with salt water covered with an oily film. A salt merchant from Pittsburgh decided to make a business of selling this as an elixir water in its natural state, undiluted, in bottles with labels that praised its potential virtues.

In 1840 a chemist from Yale University distilled the contents of a bottle. The extract was found to be light and inflammable, shedding a brilliant light. He had rediscovered the lamp oil that had been used by the peoples of antiquity. The discovery came at a critical time. The whale oil which was used for lighting was becoming scarce. Oil sources for lamps were developed from this well water, but by 1859 existing sources were no longer sufficient to meet the increasing demand.

At this time a New York lawyer, George H. Bissell, decided to exploit this mineral source. Along with James M. Townsend, a banker from New Haven, and Benjamin Silliman, a chemistry and geology professor at Yale University, he started the Pennsylvania Oil Company.

In Titusville, Pennsylvania, the group rented 125 acres of ground and one of the shareholders of the Company, Edwin L. Drake, sank the first well. Drake was an ex-railroad conductor who just happened to select the right spot to drill. At a depth of 69-1/2 feet he struck a large pool. On the advice of an old well-digger he sank an iron pipe into the ground. A small steam-engine was moved in and a wooden scaffolding installed with a chain and vertical ram. Drilling started in June 1859 and by the 27th of August oil appeared. During the next week, it flowed out at the rate of ten barrels a day.

This started a stampede and thousands of wells were drilled in Pennsylvania in the next few months. The region was quickly parceled out into many plots. The speculators dug at random. Distillers set themselves up alongside the wells and distilled the crude oil in primitive stills for the valuable lamp oil. Gas and gasoline were treated as useless.

Production rose quickly. In the State of Pennsylvania it jumped from 2,000 barrels in 1860 to 3 million in 1862. The search soon extended into neighboring states. By the end of the century oil was being produced in Ohio, West Virginia, Kansas, Texas, Oklahoma, Colorado, Wyoming, California, Russia, the Dutch East Indies and Poland.

In New York, Petroleum was quoted on the Stock Exchange. In 1862 production exceeded demand and prices collapsed. The value of a barrel (42 gallons) dropped from $20 in 1859 to 10 cents. The petroleum industry needed reorganizing and John D. Rockefeller took on the task. The risks of prospecting and extracting were left to the producers.

The first oil company was Standard Oil which centralized the operations of storage, refining and transport. The distilled products would have a uniform quality to allow stabilization of prices. Agreements were made with rail and shipping companies to transport the oil to distant areas and countries. Pipelines were also built.

As a result of these agreements, Rockefeller controlled essentially the entire oil industry of the United States by 1878. Government concern at this state within the state resulted in anti-trust proceedings against Standard Oil. In 1911 it was forced to liquidate or surrender control of 33 subsidiary companies. (See Table 2-1)

The diversity of petroleum's products, the high heat value of its fuels and its economical transportability has made petroleum, in the form of natural gases, gasoline, gas-oils and fuel-oils a major factor in the world economy.

Table 2-1. The Development of Petroleum

1860	Lamp oil used on an industrial scale
1900	Development of internal combustion engine, utilization of gasoline
1915	Manufacture of chemical products with petroleum base
1930	Use of liquefied propane and butane
1933	Development of diesel engine, utilization of gas-oil
1950	Development of jet engine, renewed utilization of paraffin

MINERAL FUELS

Mineral fuels can be divided into three types: solid, liquid and gas. In the first group are the coals. Mineral coals consist of black coal with the principal component of carbon. Some mineral coals such as anthracite are almost pure carbon. In the second group are the petroleum products which are rich in both carbon and hydrogen. These products provide a large range of fuels and lubricants. In the third group are the natural gases which often occur in petroleum deposits and the butane gases, coal, gas and the produced gases made by passing air or water vapor over hot coke.

The liquids are known as gasolines or petrols. Their physical state allows them to be used directly in spark-ignition engines. Gasoline and other liquid fuels are characterized by very rapid oxidation.

Combustion and Efficiency

The principal combustible elements common in wood, coal and petroleum are carbon and hydrogen. Of the two, hydrogen is more efficient.

The value of a fuel depends mainly in its calorific value. This is the number of British Thermal Units liberated in the combustion of one pound of the fuel. The Btu is the quantity of heat required to raise the temperature of 1 pound of water 1°F.

Pure carbon has a calorific value of 14,137 Btus while hydrogen has a value of 61,493 Btus. This means the higher the proportion of hydrogen a fuel contains, the more energy it will provide. The hydrogen content of liquid and gaseous fuels ranges from 10 to 50% by weight. They provide far more heat than solid fuels, which are 5% to 12% hydrogen by weight.

Another important consideration is the amount of oxygen. The less oxygen in the fuel, the more easily the hydrogen and carbon will burn. Thus, the lower the oxygen content of a fuel, the better it will burn. The ideal fuel would be pure hydrogen. Other factors that need to be considered in assessing the merits of different fuels are the moisture content and the ease of extraction, transportation and utilization.

Coal and petroleum seem to have organic origins. We are more certain of coal's plant-derived origin, although the exact mechanics of the process are not always agreed upon. Most believe in an organic origin for petroleum.

Most fuels come directly or indirectly from carbohydrates, vegetable matter which is the result of photosynthesis occurring in green plants. The energy in these fuels is due to the sun. When burning wood or alcohol, fuels extracted from living plants, we are recovering recent solar energy. When burning coal, or gas, we are tapping ancient solar energy.

Coke and Gas Fuels

Coking coal is a type of bituminous coal that softens into a paste as combustion takes place. At higher temperatures, its components are given off as bubbles of gas. Coke is the hard gray mass left after heating.

Bituminous coals are those richest in volatile matter and make the best coking and gas coals. In high-grade coke, the volatile content of the coal ranges from 25 to 30%. The sulphur content is higher for second-grade coke. Coking coals may yield from 50% to 80% coke.

The components of coal are separated by a process variously called distillation, carbonization, pryogenation, and coking. It differs from combustion in that it takes place in a closed vessel in the absence of air, thus preserving the calorific value of the products.

This is done in large vertical or horizontal ovens where the gas, coke and by-products are produced at high, mean or low pressure, sometimes continuously and sometimes discontinuously. Modern by-product or chemical recovery ovens save the volatile components and produce coke.

Large plants may use hundreds of horizontal ovens, in batteries of 50 to 60. A typical oven is 45 feet long by 13 feet high by 7 feet wide. It is made of steel with a lining of silica bricks and hermetically sealed by a door at each end.

In the sidewalls separating the furnaces, there are flues for boilers which operate at a temperature between 2,200°F and 2,700°F. They burn a mixture of rich gas from the ovens and poor gas from gas generators. Each oven is fed with about 15 tons of pulverized coal by an automatic coal-car. The doors are hermetically sealed and distillation begins.

The coal becomes pasty starting at the walls and proceeding to the center, it cakes into white-hot coke. The firing operation may last from 12 to 24 hours, depending on the coal used and the products to be obtained. When distillation is complete, one door of the oven is opened and the coke bar is pushed out and quenched.

In modern coking plants water-quenching has been replaced by dry-quenching, where the coke is smothered. This allows much of the heat of the coke charge to be recovered for various needs in the plant. In general low temperature processes produce a softer coke and a larger proportion of by-products.

Coke is used in blast-furnace operations, foundries and in the manufacture of water-gas, producer gas, domestic heating, and industrial purposes.

The needs of gas coking-plants differ from the needs of gas generators. The poorest quality coke is utilized in gas generators and in the manufacture of water-gas. The gas given off from the ovens is a heavy yellowish gas, with a temperature of 1800°F. Before it can be used it has to be scrubbed and cooled. The gases are sent through a series of condensers and scrubbers. Large pipes are placed above the ovens where the gas is subject to a shower of cold ammoniacal water. At this time it precipitates some of its tar. The primary and secondary condensers are used to get out the rest of the tar. Then the water from scrubbers absorbs the ammonia and the oils are distilled. Iron oxide purifiers remove toxic products such as sulphuretted hydrogen and hydrocyanic acid.

The distillation of one ton of soft coal yields about 1,500 pound of coke, 100 pounds of tar for road construction and waterproofing, 10 pounds of benzol which may be distilled to provide gasoline and napthalene, 7 pounds of ammonia for the manufacture of fertilizer and 12,000 cubic feet of gas for heating.

Coal Oil

Coal oil for lamps was already being produced when the Drake discovery well was drilled in 1859. In 1846 Abraham Gesner prepared coal oil from coal by thermal decomposition. Sperm oil was becoming expensive and by 1859 there were about 60 coal-distillation plants in the United States. Within a few years of the Drake discovery most of these plants had been converted to petroleum processing.

City Gas

The use of city or town gas was common by the end of the 18th century. Early pioneers included Jean Pierre Minkelers in Holland in 1784, William Murdock in England in 1772, and Philippe Lebon in France in 1796. Philippe Lebon foresaw a future using this gas, but war delayed his efforts to develop public and domestic gas lighting in France. Murdock in England set up a small experimental plant in 1795 and lighted a Birmingham factory with gas, receiving a medal from the Royal Society of London for his invention. The first gas-works at Westminster provided light for the streets of London in 1813. Gas lighting in the United States started in Baltimore. Other cities followed Baltimore's example.

This type of gas has the following chemical composition: hydrogen 50%, methane 25%, carbon monoxide 10% and acetylene, ethylene, sulphuretted hydrogen, ammonia, carbon disulphide, benzene, and tar droplets 15%. Tar forms a blackish deposit on walls, and most of the other components give off an unpleasant smell.

Coal Seam Distillation

The idea of distilling coal seams in situ was fostered by the Russian chemist Mendeleev (1889) and the English chemist Ramsay (1912). In 1935 an experimental station was set up in a coalfield and industrial development was started in several countries.

One method consists in firing an inclined panel of coal 650 to 1,000 feet long. It is surrounded by fire-proof corridors to prevent fire from breaking into other parts of the mine. The coal is ignited by remote control and at first the fire is fanned by cold air.

The gases derived from the distillation rise to the surface. When the reaction is under way, the temperature is raised and the proportion of oxygen in the incoming air is reduced. The distillation spreads from the bottom to the top of the coal panel.

Another process is called electro-linking, where two boreholes are drilled close to each other. An electric cable terminating in an electrode is lowered in each hole. A potential of 2,000 volts is passed through the coal.

The coal seam becomes hot, then white-hot, and finally begins to gasify. After this electro-carbonization occurs a current of fresh air is sent into one of the boreholes to collect the combustible gas.

Underground gasification eliminates much of the mining and its operational costs are relatively low. It can be used in seams which are thin or poor in quality. But the method produces no coke.

Storing the gases has been done with water gasometers and dry tanks. In petroleum-bearing areas porous rocks occupying anticlinal vaults can be used as gasometers after their contents had been removed during petroleum exploitation. In the 1960s about a quarter of the gas used in the United States was pumped each summer into storage pools and unused wells to be stored until the cold weather.

Natural Gas

In the United States natural gas has replaced manufactured gas because of improved means of transportation from remote areas. By the 1960s natural gas supplied nearly a third of the total energy requirement of the United States and was the sixth largest industry in terms of investment. The output in 1962 reached an estimated value of over 2 billion dollars and reserves had already been discovered to satisfy 20 years of consumption at the 1963 rate.

Natural gas is used for domestic purposes, and to perform many processing and manufacturing tasks. Its by-products are used in thousands of products. Many plants convert natural gas into electricity at high efficiency levels.

TAR PRODUCTS

Formerly tar was derived from coal gas and used unprocessed for tarring roads. It was found that tar contained a large number of useful products and a new industry emerged.

The tar is distilled at a constant pressure in tube-stills where the distilled product passes into fractionating columns. These are divided into a series of chambers, or vats, where the rising vapors successively deposit their components in order of increasing volatility.

The most volatile components of tar are light oils composed mainly of benzene (C_6H_6), a hydrocarbon known as the starting point of the aromatic series. Its chemical structure can be represented as a hexagon with a carbon and a hydrogen atom at each of peak. Carbon is tetravalent and each of its atoms holds a hydrogen atom with three neighboring carbon atoms.

The properties of benzene evolve from this formula, in which each of the hydrogen atoms may be replaced by another atom or by a group of monovalent atoms. Starting with benzene, chlorobenzene, nitrobenzene, aminobenzene or aniline, toluene, phenol, xylene and others are obtained. Benzene unit structures can combine to form more complex bodies, such as naphthalene and anthracene.

Benzene is useful in dry cleaning and stain removing. Paradichlorobenzene is an insecticide. Tritrinitrobenzene and trinitrotolune (tolite) are used as explosives. Aniline is a dye and the phenyles, benzyles and benzoates are used in the synthesis of perfumes.

Benzene derivatives are used in the manufacture of plastic materials and synthetic rubber (buna). Pharmaceutical products include antipyrin, pyramidon, dulcin, saccharine, luninal, stovaine, coramine, paregoric and sulfonamides.

Industrial benzol is a mixture of benzene, toulene and xylene. It can be blended with gasoline or with gasoline and alcohol. It has a calorific value greater than petroleum spirits. Its anti-knock qualities are useful in combustion engines.

Phenolics

The oils distilled from tar after the light oils are the phenolic oils. Their main components are the phenols and cresols. Pure phenol or phenolic acid is a powerful antiseptic and disinfectant. Together with the cresols, it is used in the manufacture of antiseptic soaps. It is the starting point in the synthesis of the products shown in Table 2-2.

Table 2-2. Phenolic Products

Photographic developers	hydroguinone and disminopphenol
Explosives	picric acid and melinite
Pharmaceuticals	salicyclic acid, guaiacol, menthol and aspirin
Coloring material	picric acid and fluorescein
Plastic materials	bakelite
Synthetic fabrics	nylon and imitation leathers

The middle or naphthalene oils separate in that part of the fractionating column with a temperature of 410-460°F. Products include dyes, perfumes, plastic materials, pharmaceutical and photographic materials. After the deposit of naphthalene, creosote oil is left. This has been used as a wood preservative. Heavy or anthracene oils are obtained at the bottom of the fractionating column, at temperatures between 460°F and 680°F.

The availability of many useful fuels and its economical transportability has made petroleum, in the form of natural gases, gasoline, gas-oils and fuel-oils a major factor in the world economy.

PETROLEUM

Crude oil is a substance which emerges from the ground as a thick, viscous, brown or dark green liquid. Some types like the deposits of Venezuela, contain primarily gasolines. Others, like those of Texas, are richer in fuel-oil while others like those of Mexico contain more bitumen for asphalt. The specific gravity varies and provides an indication of quality. Other criteria include pour- and cold-points, viscosity, optical properties, odor flash- and burning-points, color and coefficient of expansion.

Crude oil is the naturally occurring liquid which is a mixture of compounds of carbon and hydrogen. The proportions of the hydrocarbons vary significantly. Some of them occur as gases, others as solids, and both are in solution in liquid hydrocarbons. Since crude oil is a mixture, it does not have a fixed chemical composition or physical properties. The composition of crude oils ranges from 83 to 86% carbon, and from 11 to 14% hydrogen.

The amount of hydrocarbons in a crude oil is a measure of its purity. Generally all but 0.3 to 3.0% by weight will consist of hydrocarbons. The main hydrocarbons are paraffins, naphthenes, or aromatic groups or complexes.

Most oils have a paraffin, naphthene or mixed paraffin-naphthen base. Paraffin oil has a higher hydrogen content relative to carbon. Naphthene oil has a higher carbon content.

Naphthene oils are heavier and higher in viscous, volatile lubricating oils. Naphthene oils have an asphalt base since distillation produces solid and semi-solid asphalt residues. Lower densities and residues of petroleum or paraffin wax are a characteristic of paraffin oils.

Hydrocarbons are grouped into a number of chemical series, with different chemical and physical properties. The four series make up the bulk of crude oil are shown in Table 2-3.

Table 2-3. Crude Oil Chemical Series

Methane	normal paraffin
Isoparaffin	branched-chain paraffins
Cyclo-paraffin	naphthene series
Benzene	aromatic

The methane, or paraffin, series starts as a simple saturated, straight-chain, homologous series, advancing by a CH_2 increment from its simplest member methane (CH_2) to (CH_4) to complex molecules with more than 60 carbon atoms. A carbon atom has four arms (four valencies) which can retain four hydrogen atoms, since each hydrogen atom has only a single valency. The resulting hydrocarbon is methane with the chemical formula of which is expressed as CH_4 or alternatively as CH_3-H. This is the simplest of all the hydrocarbons, and the most stable. It is the main component of natural gas and occurs in swamps as marsh gas from decaying vegetable matter.

The formula CH_3-H contains the CH_3 group, which is called methyl. It plays an important part in the composition of the hydrocarbons. The methyl group has only one valency and replaces any one hydrogen atom of a pre-existing hydrocarbon. The first five hydrocarbons of this series are shown in Table 2-4.

These hydrocarbons are in open chain and are often written in a condensed form as CH_4, C_2H_6, C_3H_8, C_4H_{10} or C_5H_{12}. The general formula for the hydrocarbons of the methane series is $C_nH_{2n} + 2$.

Table 2-4. First Five Hydrocarbons of the Methyl Group

Methane (CH_3-H)
Ethane (CH_3-CH_3)
Propane (CH_3-CH_3-CH_3)
Butane (CH_3-CH_2-CH_2-HH_3)
Pentane (CH_3-CH_2-CH_2-CH_2-CH_3)

Most crude oils are made up of these hydrocarbons:
Gaseous hydrocarbons from CH_4 to C_4H_{10}
Liquid hydrocarbons from C_5H_{12} to $C_{15}H_{32}$
Solid hydrocarbons from $C_{16}H_{34}$ to $C_{35}H_{72}$

An oil containing only 35 hydrocarbons would be regarded as very simple. In practice, they are joined by other hydrocarbons called isomers with the same composition but differing in molecular structure and having different properties.

For example, Butane is (CH_3-CH_2-CH_3). This chain can be grafted by replacing a hydrogen atom with a methyl group and we have,

$$C_3\text{- } CH_2\text{- } CH\text{- } CH_3$$
$$|$$
$$CH_3$$

This hydrocarbon is methylbutane. Its shorter formula is CH_5C_{12}, which is the same as pentane. Methylbutane and pentane are isomers. There are three pentanes, each with the formula C_5H_{12} each containing 83.33% carbon and 16.76% hydrogen, each with a molecular weight of 72.15 but a different boiling point.

The number of isomers increases with the number of carbon atoms (n). There are

3 isomers when n = 5
5 isomers when n = 6
9 isomers when n = 7

and over 60 billion isomers when n = 40. Very few of these compounds actually occur in measurable amounts in crude oil. But, the total number of individual hydrocarbon compounds that may be present in oil is considered to be 8,000.

Crude oil is never composed of a single hydrocarbon series. The methane of the paraffin series dominates in oils of Pennsylvania, but this oil also contains hydrocarbons of the naphthene (CH_nH_{2n}) and the aromatic (CH_nH_{2n-6}) series.

Hydrocarbons of the naphthene and aromatic series differ from those of the methane or paraffin series since their carbon atoms do have all the hydrogen atoms that they could. They are unsaturated hydrocarbons, in contrast to paraffinic hydrocarbons which are completely saturated. Unsaturated carbons are not stable and tend to continuously acquire new hydrogen atoms. Other substances in crude oil include sulphur, oxygen, nitrogen and various metals.

Sulphur

The amounts of sulphur ranges from 0.1 to 5.5% by weight. It is rarely present as elemental sulphur, but usually occurs in compounds such as hydrogen sulphide (H_2S) and organic sulphur compounds. Sulphur and sulphur compounds are corrosive, emit a characteristic sulphur odor and impede combustion.

Modern refining methods eliminate most of the sulphur in crude oils. In some gas pools with a high sulphur content, the sulphur is extracted at a profit. Generally crudes of a high specific gravity contain more sulphur than others. Heavy Mexican crudes can have from 3 to 5% sulphur and are among the world's highest sulphur oils. Crudes with less than 0.5% sulphur, such as Pennsylvania oil, are among the low-sulphur crudes. About half of the oil produced in the United States is low-sulphur crude.

Most crude oils contain small quantities of nitrogen. This ranges from 0.05 to 0.08% in some California crudes.

The amount of oxygen in crude oil ranges from 0.1 to 4% by weight, averaging less than 2%. It can occur as free oxygen or in various compounds. Like sulphur, oxygen occurs in larger amounts in the heavier crudes.

Crude oil usually contains a number of miscellaneous substances, which can be organic or inorganic. There are microscopic fragments of petrified wood, spores, algae, insect scales, shell fragments and bits of lignite and coal. Many metals found in crude oil ash including vanadium, nickel, cobalt, copper, zinc, silver, lead, tin, iron, molybdenum, chromium, manganese, arsenic and uranium.

Sodium chloride is also present in most crude oils. When the quantity exceeds 15 to 25 pounds per 1,000 barrels of crude oil, desalting is required to remove the corrosive salt. Some of the salt is in the form of crystals in the oil and some is dissolved in the water which is normally produced along with the crude oil.

Distillation

Petroleum and natural gas underground and in their natural state are under greater pressures and higher temperatures than they are at the surface. As the crude oil moves upward in the well, from the reservoir to the surface, it experiences changes in pressure and temperature. This produces changes in its original composition.

Crude oil is a liquid containing gases in solution with microscopic particles of solid hydrocarbons in suspension. Because these substances do not have the same boiling point (the point where they pass into the gaseous state) or the same freezing point (the point where they pass into the solid state), it is possible to separate them as the temperature increases. This is the basic method of distillation in the refineries. At first the gases (methane to butane) are released, then the liquids are separated in order of growing complexity. The solid and semi-solid hydrocarbons form the residues of the distillation.

Since petroleum is a mixture, its physical properties vary considerably, depending upon the type and proportions of the hydrocarbons and impurities present. These physical properties include the density, viscosity, optical activity, refractive index, color, fluorescence, odor, pour- and cold-points, flash- and burning-points, coefficient of expansion, surface and interfacial tension, capillarity and absorption.

An important property is the density. The density of a substance is the weight per given volume, for example, the number of pounds per cubic foot. Specific gravity is a way of expressing the same thing without specifying a unit of measure. It is the ratio of the weight of a given volume of a substance and the weight of an equal volume of pure water at a particular temperature and pressure. In the United States, the specific gravity of oil is measured at 60°F at one atmosphere of pressure.

The price of crude oil is generally based on its specific gravity. Two scales are used: degrees Baume or degrees API (American Petroleum Institute). Degrees Baume is a European scale. In the API scale, water has a value of ten, heavier substances have numbers lower than ten and lighter substances have higher numbers.

Crude oil density ranges from types that are heavier than water. Some California crudes have an API reading of 5°-7°. Crude oils of 10°API is found in Venezuela and up to 60° API in other areas. Most crude oil averages 27° to 35° API.

The density is measured by placing a hydrometer with the API calibrations into the crude oil. Temperature and pressure readings are recorded at the same time and the reading is converted to the standard at 60°F and 1 atmosphere.

Viscosity or resistance to flow is the opposite of fluidity. It is measured by the time taken for a given amount of oil to flow through a unit opening at a given temperature. Crude oils vary greatly in viscosity. Light oils and natural gas are low in viscosity while others may be highly viscous (plastic). The viscosity depends on the amount of gas dissolved in it and on the temperature.

At ordinary temperatures some crudes are so viscous that they must be heated to be pumped. This makes production and transportation much more expensive.

The color of petroleum varies from almost clear to red, green and opaque black. Lighter colors are found in the low specific gravity crudes. All oils show some degree of fluorescence, especially the aromatic oils. Colors range from yellow through green to blue when the oil is viewed under ultraviolet light. This property is used to detect oil in drilled cores, cuttings and mud. Fluorescence allows the detection of one part in 100,000 by the naked eye, with instruments one part in hundreds of millions can be detected.

Some petroleums have a gasoline odor and those with sulphur or nitrogen have a sour smell. Table 2-5 lists some important crude oil characteristics.

Table 2-5. Crude Oil Characteristics

Cloud-point—temperature where solid paraffin waxes settle out
Pour-point—lowest temperature where oil is fluid
Flash-point—temperature where vapors will flash ignite
Burning- point—temperature where oil will burn with a steady flame

The flash and burning points are key to the hazards involved in storing and handling the oil. The others are important in transportation of the oil. Pour-points can vary from 90°F to 70°F and lower.

Petroleum Origins

Petroleum deposits usually occur in sedimentary rocks. Most petroleum in the United States occurs where there is an abundance of marine sedimentary rocks. This is also true in the regions around the Gulf of Mexico, the Persian Gulf and the Caspian Sea. Brazil contains mostly igneous rocks and is poor in petroleum. Volcanic material or igneous rocks are void of oil except in very rare cases.

Petroleum is usually related with sediments of marine origin and its water content is similar to sea water. Few petroleum deposits are found in continental sediments.

The first theories of petroleum origin suggested inorganic sources. These theories are no longer considered probable. The idea of an inorganic source came from the hydrocarbons themselves, since methane, ethane, acetylene, benzene can be made from inorganic substances.

The hydrogen content of petroleum is much higher than in most organic substances. It is possible that this hydrogen is inorganic in origin, acquired from the rocks through which the oil passes. It is also possible that the hydrogen was added by organic processes such as bacterial action. Table 2-6 lists the basic reasons for organic origin.

Table 2-6. Theory of Organic Origin of Petroleum

- Abundant and widespread source
- Organic matter with hydrogen, carbon, porphyrin pigments and nitrogen
- Association with sedimentary rock
- Optical rotary power*
- Similarity of associated substances in petroleum and organic ash
- Presence of microscopic organic fragments
- Hydrocarbons in sediments

*Except for cinnabar and quartz, inorganic substances do not have the ability to rotate the plane of polarization of polarized light. This is thought to be due to the presence of cholesterol, which is present in animals, plants and crude oil.

In marine sediments such as limestone, sand or clay, the carbon and hydrogen necessary for the formation of hydrocarbons can be supplied by living organisms from their fats, waxes and resins, which are rich in hydrogen and resist decay more than the carbohydrates and proteins.

The organisms capable of producing the fatty substances required for the formation of petroleum are thought to have accumulated in sedimentary rocks over the centuries. Some may be the microscopic animals and plants which inhabited the upper layers of the sea such as plankton.

Petroleum always contains nitrogen, sulphur and phosphorous in the form of organic compounds like porphyrin. Porphyrins are formed from the green coloring matter (chlorophyll), or from the red coloring matter (hemin) of blood. They occur in crude oil as complex hydrocarbons which oxidize readily.

In crude oil, the vegetable porphyrins are more abundant than the animal porphyrins. Bays, gulfs, coastal lagoons, enclosed seas and submarine basins with poor water circulation would enable petroleum deposits to form.

The decomposition of organic matter in the presence of abundant oxygen produces the end products of carbon dioxide and water. For hydrocarbons to be produced from the decay of organic matter the process must be stopped short of completion. The most probable areas for this are reducing zones where there is rapid sedimentation. The decaying material may have mixed with clay particles and sea water, forming a brine in where anaerobic bacteria gradually reduced the oxygen content.

In most petroleum only 4% oxygen is present, while organic matter contains about 15 to 35%. This would raise the proportion of carbon and hydrogen, which combined into hydrocarbons. Bacteria of this kind is found in marine sediments and in the salt water of petroleum containing sediments. They are essentially the same as the bacteria found in swamps that cause the formation of methane or marsh gas.

Sources and Reservoirs

Petroleum is a normal component of sedimentary rocks of marine origin and every sedimentary rock, no matter where it occurs, contains some deposits. Petroleum takes advantage of any permeable rock pointing toward the surface to rise as high as can. Hydrocarbon seepages, asphalts found on the surface of the soil, bituminous shales, bituminous limestones, and fossil paraffins are all evidence of this. They may indicate underground deposits of petroleum but often they are proof of evaporation and natural exhaustion over time.

Oil seems to migrate from fine-textured source rocks to coarser reservoir rocks with larger pore spaces. Oil migration depends on the compaction of the sediments and the relative buoyancy of oil and gas. The capillary action of water moves it into rocks with small pores and displaces any oil present into coarser adjacent rocks. The water will carry along patches of oil as it moves through the subsurface rocks.

Oil leaving a source rock tends to migrate upward in the direction of lowest pressures. This migration may be deflected or stopped by an impervious rock layer, called an oil trap or cap. Salt, shale, gypsum, and dense limestones form cap rocks. If an impervious cap is pierced into the oil, the liberated petroleum can gush out under the pressure of the confined gases. Table 2-7 lists the characteristics of oil pools.

Table 2-7. Oil Pool Characteristics

- Sedimentary rocks, predominantly marine
- Impervious rock layer (cap rock) over the oil
- Reservoir rock (porous and permeable) sandstone, limestone, contains the oil
- Reservoir trap, produced by folds, faults, variations in porosity and permeability
- Source rocks, shale or clay.

The petroleum is not free in underground cavities but enclosed within the permeable reservoir rocks. It is zoned with gas and water into layers, according to the specific gravities. The gas is usually at the top, petroleum in the middle and the salt water in the lower layer.

PETROLEUM PROSPECTING

When Edward Drake sank the first oil well as Titusville in 1859, nothing was known of prospecting methods or drilling techniques. A modest wooden structure supported a vertical ram, with an old motor and a series of iron pipes. Modern wells require a detailed study of the bedrock and can be drilled to over 20,000 feet under the most difficult conditions.

For a long time the search for petroleum was based on surface evidence of oil, the release of gases and the presence of bitumen-impregnated sand or limestone. This evidence is often misleading since it does not indicate the existence of petroleum at depth. These may be traces of former deposits now exhausted or indications of very distant deposits.

Prospecting involves aerial reconnaissance of the area to reveal faults and anticlines. Promising areas are examined by coring before final drilling.

A core drill provides a continuous rock core section by section. Analysis of the rock specimens can determine porosity, permeability, and capillarity, and indicate the potential productivity.

Core holes can be drilled to depth of more than a thousand feet. A slim-hole rig of 6-7 inches in diameter can yield a continuous core at a smaller cost than a conventional coring rig. It relative cost has resulted in its application in drilling for production wells in some areas.

Drilling

Investigation of an area can last for several years before drilling. If drilling should strike a potentially profitable pool, the chances of striking crude oil are only one in three. The drill could strike the gas layer above the oil or the salt water below.

The first step in sinking a well is to erect a steel framework or derrick to a height of 100 to 200 feet. The average height in the United States is about 180 feet. This supports the pulleys which handle the drill pipes.

Early wells were sunk by ramming. This was replaced by rotary drilling, where the bit turns at 50 to 300 revolutions a minute. Different bits are used for soft or hard rocks or coring. They tend to crumble the rock. Drilling speed can vary from 8 inches an hour in hard rocks to 100 feet per hour in soft ground.

A continuous stream of mud is pumped into the hollow drill pipe and bit while the rotary table turns. Openings at the end of the bit allow the mud to rise between the drill pipe and the sides of the well. This fluid lubricates and cools the bit while transporting excavated material to the surface. It also acts as an eroding agent in soft rocks and cakes the sides of the well to prevent drilling fluid from being lost.

Bentonite or other clays are added to the mud if there are not enough clays or solids present. Fibrous materials such as special cements, sawdust, beet pulp, cellophane and shredded plastic foil may be used as a seal. High-density fluids are used to prevent blowouts, and weighting materials, such as barite, may be added to increase the weight of the mud for deep drilling.

Water base muds are the most common but oil base muds are used for low-pressure reservoir drilling where a weighted fluid is not needed.

Oil base fluids are also used in drilling through clays, salt and anhydrite. Rotary drilling may use also air or gas instead of drilling fluid.

On its return to the surface the drilling fluid passes into a tank for sifting. Analysis of the cuttings provides information on the excavated material and possible hydrocarbons.

The bit will be replaced often depending on the hardness of the rock. Replacement may be required two or three times a day. This means drawing up the set of pipes, three or four sections at a time, and unscrewing each until the drill bit emerges. The set is reassembled and lowered. The pump is restarted and injects mud until it flows into the sifting tank. Then the drilling engines are started and the rotary table begins to revolve again starting the drilling bit on its way. Rotary drills have a speed in soft rocks of over 2,500 feet per day.

Turbo-drilling was pioneered in California in the 1940s. But, it was perfected by the Russians and it is almost standard equipment in Russia. The turbo-drill can be 12 times faster than a rotary. The bit is rotated by the hydraulic power of the mud stream pumped down through a non-rotary pipe. A turbine below the drill pipe converts the hydraulic power of the mud stream into rotary power.

After drilling an empty string of pipe with a valve is lowered. It stops short of the oil formation. When the valve is opened, the reservoir fluid flows up the drill pipe. If the flow is slow, exploding nitroglycerine, acidizing and hydraulic fracturing may be used to increase the flow. These techniques can convert a dry hole into a producing well.

Turning the test well into a commercial well involves running a casing in and cementing it in place. Inside the casing is a smaller diameter tube, open at the bottom for the oil. Other methods include a gravel pack to prevent collapse of the walls or the use a plastic cement in sand.

Drilled wells date from 221 B.C., in China, a 450 feet well was drilled for brine. But, the data and the depth of the first is unknown. In the United States wells over 2,000 feet deep were drilled by 1854. A 25,340 foot well was drilled in 1958 but was dry. One productive well of 21,443 feet is in Louisiana.

Underwater Drilling

In deep drilling, robot drilling rigs are often used where the operator simply manipulates the rig at a control console. Techniques have been developed for diverting the direction of holes from the vertical. This is used for deposits situated under built-up areas or under the sea.

Directional drilling is started vertically. At a depth of several hundred feet, a new drilling bit with a whipstock, (which is a long, slender, steel wedge) is lowered into the hole. The wedge deflects the bit from its original direction to start a new hole at an angle. Drilling with the normal bit is resumed for several hundred feet. Then the whipstock is lowered again to deflect the bit. The process is continued until the desired depth is reached.

Directional drilling allows the use of one surface well and several others that radiate from the bottom of the vertical section. This means a single drill site can serve a number of wells in offshore operations or in rugged terrain.

Directional drilling is used extensively for underwater deposits. The derricks are mounted on modified ships or erected on platforms constructed at sea.

A typical oil island is several miles off the coast. A steel platform rests on foundations 50 feet high. It must withstand winds of 160 miles per hour and waves 30 feet high. The structure is more than 170 feet high and in addition to the derrick and other equipment the island must accommodate some 50 workers. It has fresh water and other provisions with a generating station.

In offshore drilling, the equipment allows drilling from a floating vessel and the completion of oil wells on the ocean floor is accomplished by remote control from the surface. The drilling vessel uses an automatic pilot to keep it in position while drilling.

Oil Flow

Water and gas move through rocks with greater ease than oil. When the pressure at the bottom of a well is lowered by allowing the oil to flow out rapidly, the overlying gas moves down and the brine below moves up, before the oil from the pool moves in the well. The water beneath may rise up and seal off segments of the pool. The well then produces more water or gas than oil.

In the early days of drilling, wells flowed freely until pressures were reduced. Today, gushers are rare. The oil flow is controlled by a choke valve at the top of the well and the gas and water press evenly on the oil to produce a flow from the sides of the pool to the well.

Oil movement depends on three different types of forces, or drives. Oil always contains some dissolved gas. Since it is compressed it will try to expand and force the oil with it in the direction of the pressure release, toward the well hole. Oil recovery is low when dissolved gas is the only drive, since the oil-gas ratio cannot be controlled and the pressure drops quickly. The recovery may be less than 20%.

In most wells dissolved gas is supplemented by free gas which collects above the oil. This compressed gas cap expands into the porous oil rock as pressure is eased and drives the oil into the wells located near the bottom of the oil layer. The recovery in a gas-cap field is 40% to 50%.

Under the oil there is usually a large quantity of salt water and as pressure is released it moves into the oil rock and flushes the oil out. This is the most efficient of the natural drives and if the oil flushes out the edges of the pool as well as the center, recovery under favorable conditions can approach 80%.

In an oil column several thousand feet long gravity drive can be important. In these fields, almost all the recoverable oil can be eventually brought to the surface.

Some oils are too heavy and viscous to flow into wells and must be heated below ground to increase fluidity. This is done by lowering a heating unit into the well. If it increases the flow by five or more barrels per day it is considered successful. The heating can be done electrically or circulating hot water.

Sometimes heating in situ is done by the hot foot method. The oil is ignited at the bottom of one well and the fire is maintained by forcing compressed air through the hole. This heat drives the oil through the rocks into the surrounding wells.

Clogging by wax or asphalt can block the oil flow. New channels may be opened by controlled explosions of nitroglycerine or by pumping in strong acids.

Secondary Recovery

After a field ceases to produce by natural flow and by pumping, it can be repressured by injecting air or natural gas from which the gasoline content has been removed. The field may also be artificially flooded by pumping water down old wells into the reservoir. This will flush some of the remaining oil into the other wells. Acidizing involves pumping acid with an additive to deter corrosion into the reservoir.

Hydraulic fracturing may combine water, acid or crude oil with sand at high pressures to produce new cracks and enlarge existing ones in the reservoir rock. As this fluid is removed, it leaves the sand behind to prop open these fractures. In the Permian Oil Basin of the United States, fracturing operations have increased the known recoverable oil reserve of the area by 120 million barrels.

Oil and gas reach the surface as a frothy mixture which is sent to a separation tank. The oil then is pumped into storage tanks and the gas is treated in natural gasoline plants for the removal of any remaining liquids. The dry gas is used in several ways. Some is used to supply power in the oil field or pumped back to maintain pressure in the reservoir. The balance is sold or burned off. Natural gas recovery has increased importance since it is used extensively in the manufacture of chemicals and as a source of gas in the home.

Along with natural gas, oil fields have other products. Hydrogen sulphide is the most abundant of the impurities. A small quantity in natural gas is desirable as a warning of leakage. When large quantities are present this sour gas is used to manufacture carbon black or is processed for sulphur recovery. Sometimes helium is present in useful quantities. Carbon dioxide is found in some fields. It is recovered and used in the manufacture of dry ice.

Salt water is produced in vast quantities, especially as an oil field becomes exhausted. The corrosive action on pipes and fittings becomes a problem. In some pools sand comes up with the oil and must be disposed of.

The oil pool must be large enough to make a profit. Small pools are often turned down by large companies. The small prospects may be workable by individual operators whose requirements and expenses are smaller.

Crude Oil Pricing

Usually the entire income of a field is derived from the crude oil. Costs incurred up to the time when the oil enters the pipeline must be subtracted from the sale price to calculate the profit. The price at the well varies with demand, the type of oil, distance to be transported and type of transport.

A prospect near a company's own pipelines is more desirable than a remote prospect. Prospects in a heavy oil area are less desirable than those in a light oil area especially in areas with a high demand for gasoline.

Exploration varies in direct response to the price of oil and gas. As the price rises, smaller prospects and prospects with doubtful evidence or high lease costs begin to look attractive.

The cost of locating a prospect may fluctuate greatly. Sometimes the necessary data can be found in public records or government surveys and the costs are small. When an area has to be explored it may involve years. Seismic studies often result in the location of only a few questionable prospects.

Lease costs include the initial cost of the lease, a royalty on the oil produced or a proportion of oil set aside for the owner and the rental for maintaining the lease.

Drilling costs depends on the depth to be drilled as well as the nature of the rocks (hard or soft). Other factors are the distance from water and other supplies, method of transport and access to the site. In foreign countries royalties and other profit-sharing expenses or taxes may deplete 50% of the net profit and more.

Production costs are affected by undesirable factors, such as loose sand in the oil and large amounts of hydrogen sulphide gas. These will damage equipment and require costly repairs. They may cause a prospect to be abandoned. The natural drive is important along with the probable recoverable reserve.

PETROLEUM REFINING

Petroleum is a complex mixture that can serve many purposes. In order to make it useful it is necessary to separate it into its numerous components and to fractionate it into groups of hydrocarbons having certain average properties suitable for the desired purposes. Refining achieves this fractional distillation and purifies the products for consumption.

The light gases such as methane and ethane are used as fuels in the refineries and may be burnt as waste products if not saleable. Propane and butane are denser. They are first scrubbed and then liquefied under pressure and stored in tanks. Butylene is produced in various refining processes. It is the raw material used in the synthesis of rubber.

Motor fuels are an important product. Gasoline is required for spark-ignition engines. The different needs of aircraft and vehicles are constrained by the different qualities of the fuels produced in the refineries. Aviation fuels demand a high octane. Other products are added to motor fuels to improve their performance in automobiles.

Special oils are obtained after the motor fuels during distillation. These oils are used for lighter fuel, dry cleaning, solvents for cleaning, wax, polish, solvents for paint and solvents for rubber.

Kerosene or lamp oil was used in lamps and oil stoves. Today a form of it is produced for jet engines. It is also a component of insecticides and fungicides.

Gas oils have a brownish hue and are slow to evaporate. Diesel engines have made them important as fuels for trucks, tractors, locomotives and boats. They are also cracked to obtain gasoline.

Lubricating oils are used for all types of engines and machine tools. A wide range of lubricating greases and oil for printing ink are also made from them. There are also many secondary derivatives used in articles manufactured from special fuels and oils. Many products are specially refined and deodorized, such as oils for cosmetics, rust preventives and cable coatings. Fuel oils are made from distillation residues. They are used in industrial furnaces and central heating boilers.

Petroleum waxes are the white products such as paraffin and Vaseline extracted from distillation residues. They are used in candles, wax, polishes, chewing gum, pharmaceutical and waterproofing products.

Petroleum coke is a by-product of refining. It can be obtained in two forms, as thermal or green coke which is used as low ash fuel, and in a calcined form which is used in the aluminum and steel industries as well as in the nuclear industry.

Asphalts are viscous or semi-solid residues. Their adhesive and impermeable qualities make them suitable for paving roads, coating roofs, and impregnating wood. They are also used for electrical insulators.

Carbon blacks are the final residues from distillation. These carbonaceous substances have been used in the manufacture of printing inks and photograph records. They can be added to rubber to prolong its life. A tire that is not reinforced with carbon black has a life of only a few thousand miles. A tire containing carbon black can last well over 30,000 miles.

Refineries

Although petroleum was reported to be distilled in Russia in 1735, the first true refinery was built at Titusville, PA, in 1860 at a cost of $15,000. By 1873, hundreds of small refineries were in operation along with a few large ones.

The earliest refineries operated in a batch operation. Oil was poured into a vessel and the products were distilled, one by one, by increasing the temperature until a residue of heavy lubricating oil or tar remained.

The advantage of continuous processing was realized with the completion of the first Trumble plant in Vernon, California, in 1912. Modern plants are completely continuous with several distillation units connected together so that the hot product from one unit, the residue, is pumped directly to the next stage. Several units may have a common heat exchange system where the charging stock for one unit is heated in other units.

Many precautions are taken at refineries to keep out unauthorized visitors. Even authorized visitors are checked to make sure that they carry neither lighter nor matches. A typical refinery may contain hundreds of miles of pipelines with laboratories and numerous instruments which give remote readings of the processes taking place.

A modern refinery extends over several acres and is divided into sectors or units. Each is involved in a particular process or the storage of a certain product. There may be fractional distillation, reforming, cracking, rectifying, scrubbing, and transforming units.

In atmospheric or straight-run distillation the crude oil is first pumped into the fractional distillation unit. This is the refinery's tallest unit and some of its columns are used for atmospheric distillation while others are for vacuum distillation. Heated to about 680°F in the gas furnaces, the petroleum reaches the first atmospheric column, which is divided into compartments for fractional distillation. The lighter and more volatile hydrocarbons rise to the upper part. Those that are heavier and less volatile collect in the lower part. While rising, a volatile mass tends to shed its less volatile elements.

Gases and gasolines are recovered from the top of the first column. The residue is taken from the lower part, reheated and injected into a second atmospheric column for the extraction of naphtha, kerosene and part of the gas oil.

The new residue at the bottom of the second column is a mixture of gas oil, fuel oil and bitumen. This is distilled under vacuum in a large fractionating column after reheating to a temperature of 680°F. Under reduced pressure the hydrocarbons dilate and separate more completely than under atmospheric pressure.

Gas-oils and a residue rich in asphalts are obtained. The following petroleum derivatives have been separated: gases, gasolines, naphthas, kerosene, gas oils, fuel oils and asphalts. These are all in impure forms. The gases still contain gasoline, and the gasolines contain gases. The naphthas contains gasoline and kerosene. A small amount of gasoline is recovered. These products are sent through rectifying and filtering units.

The gases given off from the top of the first atmospheric column or recovered during succeeding operations are processed to eliminate methane, ethane and gasoline obtaining propane and butane. This is done with steam-heated rectifying columns for fractionation.

After filtration the propane and butane are liquefied under pressure and stored in tanks. The propane is pressure stabilized at 2 to 4 pounds per square inch and the butane at pressures of 3/4 to 1-3/4 per square inch.

Several problems arise in connection with the gasolines. To meet the heavy demand by simple distillation would require such large quantities of crude oil that the market would be saturated with the accompanying overproduction of middle and heavy products.

But, the gasolines derived from crude oil (straight-run gasoline) and those obtained by filtration of the gases are poor quality, low-octane spirits. It is possible to use them in spark-ignition utility engines, but they are unsuitable for automotive engines and aircraft. Under heavy pressure in the cylinders their combustion is irregular. Peroxides form and explode, causing a knocking effect. The power developed by these fuels is smaller, and their action is harmful to engine parts. Knocking can be prevented by adding hydrocarbons of the aromatic series, lead tetratehyl, or octane.

Cracking

Cracking or thermal decomposition was practiced before the era of petroleum in the distillation of coal and oil shale. In the early years of the petroleum industry, a cracking distillation along with the common distillation with steam was used.

Cracking distillation was accidentally discovered in 1861. A distillation was half complete, but there was a strong fire made and it was untended for 4 hours. A light-colored distillate with a low specific gravity resulted and a heavy oil was condensing on the cooler parts. The temperature was high enough to cause decomposition of the heavy oil into lower boiling point products.

In the early days of the industry, gasoline was of little value. Cracking was used to produce more kerosene than could be obtained by steam or simple distillation. At first, cracking was done by the decomposition of vapors, later liquid-phase cracking processes would include viscosity breaking and reforming.

In viscosity breaking only a mild decomposition take place, the oil is decomposed just enough to lower the viscosity and pour point, so it can be pumped more easily. Little or no gasoline is produced.

Reforming can involve a product like naphtha or low octane gasoline. These are decomposed into a higher percentage of high octane anti-knock gasoline.

Hydrogenation

Hydrogenation is similar to cracking, but hydrogen is introduced, in the presence of a catalyst, into the decomposition process. The assimilation of hydrogen occurs at high temperatures and pressures. The process can be used in the manufacture of most commercial oil products while cracking is used extensively for the production of gasoline.

Kerosene, gas oils and full oils can be partly cracked into gasolines, where their heavy hydrocarbon molecules are broken down into light hydrocarbon molecules. This is done with coil furnaces, at a temperature above 750°F. Then at a pressure of 10 to 17 pounds per square inch they are sent into reaction chambers, where the molecules are broken down. The gasoline yield from this thermal cracking is 55-70% for kerosene, 40-55% for gas oils, 30-40% for fuel oils. The residues are bitumens and carbonaceous products similar to coke.

Fractionation

Fractionation involves the separation of a liquid into products with a shorter boiling range using vaporization. In early refineries, this was done with a series of distillations where partly separated products were redistilled several times until the desired product was achieved. Next came fractionation by partial condensation. Here a mixture of vapor is condensed in portions by cooling the vapor to lower and lower temperatures.

Next came the use of pipestills and bubble towers where the vapor in the tower bubbles through the liquid on the plates in the tower. Bubble towers allow fractionation to take place while a mixture of rising vapor is scrubbed by a stream of falling oil. Bubble towers are used in topping plants, rerun operations, cracking plants and natural gas stabilization.

Natural Gasoline

Natural gasoline attracted attention in 1912 when the demand for gasoline increased. By 1936 more than 7 percent of all gasoline was produced from natural gas. The first method of recovery was compression. As a gas is compressed, its dew point is raised so upon cooling to the original temperature, a mixture of hydrocarbons condenses. This product contains a large percentage of volatile hydrocarbons, such as propane, which need to be removed before the gasoline is suitable as a motor fuel.

The first method used to remove these hydrocarbons was called weathering. This was done by allowing the gasoline to stand in an open vessel for a period of time. In a later method, the natural gas would be passed through absorption chambers (packed columns) where the gasoline was absorbed by naphtha. This evolved into an absorption process where the gasoline is absorbed by a low boiling point gas oil by heating it and stripping the gasoline from it with steam.

A process using adsorption was used from about 1920 to 1935. This involved adsorption of the gasoline by charcoal and recovery from the charcoal by steaming. High pressure absorption is a newer process for gases which are at high pressure in the field.

Catalytic cracking uses a catalyst to assist and control the cracking process. All of these processes allowed the production of 100 octane aircraft fuel and inhibiting agents were used in cracked gasoline to prevent gum formation.

Dewaxing

The filter press method of dewaxing was first used in Scotland for shale oil before the era of petroleum. It still may be used for lighter wax bearing oils.

Heavy residual wax bearing stocks were first dewaxed by cold settling. Early refiners found that a waxy layer of oil was deposited on the bottoms of crude oil storage tanks if the tanks were allowed to stand full of oil through the winter. Cold settling evolved from this method, the oil was chilled and allowed to stand in insulated tanks. The centrifuge process of dewaxing uses warm water injection to remove the petroleum wax from the centrifuge bowl.

Reforming

While cracking produces gasoline from distillation products which do not originally contain any, reforming only increases the octane of gasolines. The starting point for the process is generally supplied by the naphtha products obtained in atmospheric distillation. The temperatures and pressures required are higher than those used for cracking.

From the straight-run, cracked and reformed gasolines the refiners compound the different mixtures sold as proprietary and regional brands to meet pollution standards.

Even greater quantities of gasoline can be obtained by cracking the heavy products and also increasing the octane producing isobutane, iso-pentane, isoheptane and iso-octane in branched-chain types and in the cyclic group. Both the isos and cyclos products are superb for aircraft engines.

Cracking of this kind can only take place in the presence of catalysts. These are substances that are capable of starting chemical actions without themselves undergoing appreciable change. The process was developed in 1933 by a French engineer, Houdry. It allowed the huge production of aviation fuels needed during the Second World War.

The method starts with heating to 900°F and circulation in reaction chambers where the catalyst of aluminum silicate is disposed in an even layer on the surfaces touched by the vapors. Carbon settles on the catalysts and must be burned in contact with air for the product to be regenerated.

Filtration

In the products derived by direct distillation, reforming or cracking, there are impurities which make them unsuitable for consumer use. These must be eliminated before storage, and a section of the refinery is used for filtration. At this stage the sulphuretted hydrogen and sulphur compounds (mercaptans) are removed.

The products obtained by vacuum distillation include fuel oils and asphalts. From them the refiner extracts lubricating oils, petroleum and paraffin.

The oils are discolorized by filtration in columns of activated earth. Paraffin is removed by solvents such as methl-ethyl-ketone, (MEK) followed by chilling and by centrifuging.

Petroleum Distribution

Since petroleum is a liquid, it has always been possible to transport it by pipeline. In 1865 a pipeline, 2 inches in diameter and 5 miles long, was built between Oil Creek and Kittaning for the Pennsylvania petroleum. It carried 800 barrels a day. In 1879 the first long-distance pipeline was constructed, between Corryville and Williamsport, Pennsylvania. Later, it was continued to Bayonne, New Jersey. It had a 6 inch diameter and could deliver 10,000 barrels a day.

Today, there is a complex system of pipelines, tank-ships, barges, railroad tank-cars and tanker trucks, distributing crude oil to refiners and the finished products to distributors, supply points, service stations, airports and homes.

A main part of this network is the pipeline. Steel pipes up to a yard in diameter run for thousands of miles through densely populated areas, in America, or across deserts, in the Middle East. Powerful machines allow several miles of pipelines to be laid through the desert in a single day.

After grading the route, a ditching machine gauges out the trench for the pipeline. The sections of pipe are aligned and welded. The pipes are coated with cold and hot enamel and wrapped in tar-lined paper. Tractor-mounted derricks lift the structure in 20- to 40-foot sections and lower it into the trench. Most pipelines are buried, but some are left at surface level.

Oil does not flow through the pipeline on its own. Slight differences in level and the viscosity of the product make natural flow difficult. Pump stations must then be installed at intervals along the route. In level areas the stations are usually 35 to 75 miles apart. In hilly regions they can be much closer.

Crude oil and finished products are sent through different pipelines. A kerosene buffer is used between certain finished products to prevent mixing.

Tracing a particular shipment from a refinery to its delivery point involves knowing the speed of flow and point-to-point distance from pumping station data. The travel time is then calculated. Shortly before the expected arrival time, samples are drawn and tested. In the case of crude oil the specific gravity is checked. This continues until the expected product arrives. A shipment of 250,000 barrels represents a stream over 300 miles long in a 16-inch pipe and takes 4 days and nights of pumping to deliver.

Pipelines gradually acquire a sludge coating which is periodically cleaned off by a go-devil device that uses whirling arms or flanges to scrape the inside of the pipe. Pipeline walkers monitor the pipeline checking for leaks. Low-flying aircraft are also used to check for breaks.

Major pipelines are found in the Middle East, the United States and Russia. One pipeline runs across the Arabian desert, a distance of over 1,000 miles. In the United States, over 400,000 miles of pipeline are used to link the petroleum areas of the southwest and the industrial regions of the northeast and northwest. Finished product lines are a more recent development. They make up only a small part of the network.

When oil tankers have tonnages that prevent them from entering certain ports, they anchor at sea and are loaded and unloaded by sea lines which are pipes running out to them.

Oil Supplies

World crude oil reserves were estimated to be nearly 700 billion barrels in 1987. More than half (57%) were found in the Middle East. The countries with the largest reserve totals, in order are shown in Table 2-8.

Table 2-8. Oil Reserves

1. Saudi Arabia
2. Kuwait
3. Soviet Union
4. Mexico
5. Iran
6. Iraq

These six countries account for almost 70% of the world's crude oil reserves. The U.S. ranked eighth in world crude oil reserves, accounting for about 4% of the world total with about 28 billion barrels. Since U.S. consumption is about 2.8 billion barrels of oil annually, if the U.S. did not import foreign oil, these reserves could be exhausted in about 10 years.

A great many numbers are used in calculations of energy reserves, and there is always the question of which numbers are correct. In the past, it was a common practice for energy and natural resource analysts to tabulate all of the known oil reserves and divide the sum by the current rate of consumption to determine how long the resource would last. Using this method has resulted in many inaccurate forecasts.

The move to fuel cells may not be pushed by declining oil supplies. The cost of developing new oil discoveries continues to fall and we may not see a forced drop in productivity. It was thought that there was 1.5 billion barrels of oil in the North Sea, but now there appears to be 6 billion barrels. We may not reach the physical limits of oil production until 2050. The U.S. has essentially replaced fuel oil with natural gas in industrial consumption and electric power generation.

One estimate places U.S. reserves at 15 years, while Russia has 20 years left, Iraq 160, Kuwait 115, Saudi Arabia 80 and Venezuela 80. These estimates are usually based on present production rates and efficiency.

Reserve estimates are not always reliable since they are often inflated however in the last 25 years oil reserves have been growing. At the end of 2004 the world had about 40 years of reserves at current production levels, but at the end of 1984 this number was only 29 years. Technology has allowed the oil industry to find and produce more oil in mature regions while opening up remote and often hostile oil producing areas. At the end of 2004 proven reserves of oil were 64% higher than they were in 1984.

One half of the oil consumed in the world is used for transportation. In the U.S. 2/3 of the oil used goes to fuel transportation needs. This makes the transportation sector the focus of all our fuel needs.

The 300 million autos in the U.S. drive this transportation sector which causes the U.S. to import more than 60% of its oil. The world is predicted to have almost 4 billion vehicles by 2050. The U.S. Fuel Cell Council predicts that world oil demand will exceed 82 million barrels a year. The U.S. Energy Information Agency projects a demand of 121 million barrels a day by 2025. This could exceed $3 trillion with oil at $75 a barrel.

Renewables supply about 15% of the world's energy. Much of this is hydroelectric power. Wind is used in over 65 countries, but wind and solar are expected to provide only 1% of the world's energy by 2030. The International Energy Agency estimates that the world will need to invest $16 trillion over the next 30 years to maintain and expand the energy supply.

References

Bradshaw, MS, R, and Mazry M. Owen, Translators, *The New Larousse Encyclopedia of the Earth*, Hamlyn Publishing Group Limited: London, England 1972.

National Geographic: A Special Report on Energy, National Geographic Society,: Washington, D.C., February 1981.

CHAPTER 3

FUEL AND AUTOS

As recently as 1978, gasoline sold for less per gallon, in constant dollars, than it had in 1960. Now prices tick ever upward as reserves grow shorter and supplies less secure.

Before the 1970s, the typical American car was overweight at 22 times heavier than a 150-pound driver, overpowered, oversize, and very thirsty. In the late 1970s standard American cars were downsized by trimming weight and exterior dimensions. These redesigns produced smaller and lighter vehicles with only half the cylinders of the once dominant V-8 engine.

Front-wheel drive eliminated the shaft from transmission to rear differential and saved weight. More weight saving in such areas as bumpers, hood, and body panels took place by substituting plastics and aluminum for steel. Overdesigned frames were replaced by integrated frame-body shells similar to aircraft fuselages.

Although smaller cars use less with petroleum, they are not necessarily more efficient. Of the energy released in combustion only 12-15% is finally applied to move the car. Most of the rest is lost due to the thermodynamic inefficiency of the engine and escapes as heat. The remainder is drained off by such factors as aerodynamic drag, rolling resistance of tires, transmission slippage, internal friction, idling, and air conditioning.

Just to push the air out of its way, a car may use 50% of the available energy at 55 mph and 70% at 70 mph. Large frontal areas create air turbulence and drag. Bodies derived from wind tunnel testing can provide a more smooth air flow around the vehicle. Details such as mirrors, rain gutters, trim, wheel wells and covers can be more appropriate.

Radial tires can reduce fuel consumption as much as 3%. Punctureproof tires of plastic could save even more and eliminate the cost and weight of a spare tire and wheel.

Automatic transmissions inflict a mileage penalty of about 10% compared to manual gearboxes. Continuously variable transmissions can provide better mileage. A stop-start engine that shuts down if a car is idling or coasting can cut gas consumption by about 15%. A touch on the accelerator restarts the engine.

Better lubricants and bearings can reduce friction, and microprocessors can monitor systems and make adjustments to keep operation at peak efficiency without or despite actions by the driver.

As the price of petroleum for gasoline and diesel engines converges with that of alternate energy sources, new power systems will become widely used. Battery-powered electric motors have their advantages; quietness, low pollution and simplicity. But, their disadvantages of limited range between recharges (which are also limited), weight, and bulk reduce their market potential. New battery systems could give better performance but they have not been forthcoming. Efficiency can also be increased by using flywheels to equalize power demands on batteries during acceleration and hill climbing.

Electric motors are paired with small combustion engines in hybrid systems with electric power for low speeds and combustion for highway passing.

Power systems that run on compressed gases such as propane, methane, or hydrogen are problematical. Range may be limited since distribution systems are not in place and each station pump could cost $30,000.

So-called synthetic fuels could be used directly in engines or to generate electricity from fuel cells for electric motors. Other combustion engines such as the sterling motor may also become options.

The proper maintenance of roads can improve mileage 5%. Since combustion engines operate best at about 45 MPH, traffic ideally should be speeded up in cities and slowed down in the country. This will prove to be difficult though possible, low highway speeds are unpopular and higher city speeds are impractical for safety reasons.

RISE OF THE AUTO

The rise of the automobile is a well-known part of the American past. The first automobiles in the 1890s were marketed as a luxury item for the rich, but the car would become a mass-produced commodity in a few decades.

We often think of the automobile as an invention that closed the 19th century and was welcomed in the 20th. In 1894, Frank and Charles Duryea of Springfield, MA, were receiving orders for gasoline buggies. This new industry would come to influence many facets of modern life from housing to recreation.

In an earlier time, hundreds of years before the Duryea brothers, Leonardo da Vinci considered carriages that would move under their own power in the 15th century. He made drawings showing steering and transmission systems. Other Renaissance painters left sketches of carriages that were propelled by geared cranks.

Steam Power

The Flemish Jesuit priest and astronomer Ferdinand Verbiest, who mapped the early boundary lines between Russia and China also made a miniature four-wheeled steam carriage for the Chinese Emperor Khang Hsi in the era between 1665 and 1680. Detailed plans of this two-foot-long carriage have been found and working models have been constructed from them.

A workable steam car became a reality in 1769 when Nicholas Cugnot, a military engineer designed a steam-powered military truck that was capable of 6 miles per hour. A second and larger model had front-wheel drive with the boiler in front of the drive. In England, steam carriages were quickly adopted into public transportation.

James Watt's technical innovations made the Age of Steam possible. He was granted a patent for a steam carriage in 1786 although he was afraid of explosions from high-pressure boilers and would not allow steam carriages near his home.

Richard Trevithick was another pioneer of steam engines. He developed and patented a locomotive-like carriage with a boiler and smokestack in 1801. Trevithick's car had eight foot rear wheels.

Another British inventor, Goldsworthy Gurney built a steam car in 1825, that made an 85 mile round trip in 10 hours. It was later damaged by anti-machinery Luddites and Gurney and his assistant engineer were knocked unconscious.

In America, Oliver Evans was an early 19th century engineer who built a self-propelled vehicle that also ran in water. Evans built this 20-ton unit in 1805 as a dredger to excavate the waterfront in Philadelphia. The Schuylkill River was a mile from his workshop, so he drove it up Market Street at 4 miles per hour. America was developing its rail system and by 1850, there were 9,000 miles of railroads in the U.S.

In England, Walter Hancock was the first to offer regular passenger routes. From 1824 and 1836, Hancock had nine steam coaches with paying customers.

Steam carriages were used as fire engines in several cities after the Civil War was over. Steam cars competed with internal combustion vehicles in the first decade of the 20th century. The Stanley, Locomobile, and White steam cars had many fans. The cars were admired for their silent option, range and there was no need of a crank handle for starting. Starter motors would appear much later. This edge was shared by electric cars and was a real consideration since careless cranking could injure your arm.

However, the steam cars needed up to half an hour to build up a head of steam and required large amounts of water and wood. There was also the fear of reported boiler explosions. The gasoline engine was greatly improved after 1905 and the use of steam cars disappeared. By 1911, White and Locomobile discarded steamers and switched to gasoline engines.

Early Electric Cars

Electric cars would be popular for a while but they eventually lose out to the wider-range of the gasoline car. The lack of good roads outside the cities forced most of the early traffic be local. Until American roads improved, almost all cars kept within the city limits where the short range of the electric car was not a problem.

The United States had almost 30,000 miles of roads by the 1830s, but most were dirt horse tracks. By 1905, both gas and electric cars were appearing, but only 10% of U.S. roads were paved. Many of the early automobiles were high wheelers to handle the heavy mud that formed when it rained. Horse traffic turned streets into mud pools.

The early electric cars were favored by many city dwellers. Electric vehicles grew out of experiments with electric trains and trolleys. They became practical with the invention of the storage battery in 1859.

A three-wheeled electric carriage was made by Magnus Volk of Brighton, England, in 1888. One of the earliest electrics was a four-passenger carriage, with a one-horsepower motor and 24 cell battery, built by Immisch & Company for the Sultan of Turkey, also in 1888. William Morrison of Des Moines, Iowa, appeared at the World's Columbian Exhibition in Chicago in 1893 with a six passenger electric wagon.

Many thought that the self-propelled automobile was a novelty with no immediate practical use, since trains ran on the only usable roadbeds and trolleys with 850 systems in the United States by 1895 were the transportation in the cities.

Electric cars made some impact through their use as taxis. In the New York of 1898, the Electric Carriage and Wagon Company had a fleet of 12 electric cabs with well-appointed interiors available on the city streets. They resembled contemporary horse carriages since the driver sat outside on a raised platform.

These small, successful businesses allowed electric cars to be accepted. In 1900, at the first National Automobile Show a poll showed that electric power was the first choice, followed closely by steam. Gasoline was a distant third, with only 5% of the vote. During that year almost 1,700 steam, 1,600 electric and 100 gasoline cars were made.

Many steam car developers did not get beyond building a few hundred or thousands units, but many of the early gasoline car pioneers became major manufacturers. Gottlieb Daimler, Henry Ford, Ransom Olds, Carl Benz, William Durant (General Motors founder), James Packard and John Studebaker are a few who played important roles in the early years of autos.

Some of these began their work with electric cars. In Germany Ferdinand Porsche, built his first car, the Lohner Electric Chaise, in 1898 at the age of 23. The Lohner-Porsche was a first front-wheel drive car with four-wheel brakes and an automatic transmission. It used one electric motor in each of the four wheel hubs similar to today's hybrid cars, which have both gas and electric power.

Porsche's second car was a hybrid, with an internal-combustion engine driving a generator to power the electric motors in the wheel hubs. On battery power alone, the car could travel 38 miles.

One device that hurt the electric car was the self-starter for gasoline engines. It was first marketed to women. Cranking from the seat instead of the street would eliminate a major advantage of the early electric cars.

Charles Kettering's starter caught on quickly and the sales of electrics dropped to 6,000 vehicles, only 1% of the total, by 1913. In that year, sales of the Ford Model T alone were over 180,000. Electric carmakers closed down or united. There were almost 30 companies selling electrics in 1910 and less than 10 at the end of World War I. A few, such as the industry leader Detroit Electric, lasted into the 1920s.

Another early hybrid was the Woods Dual Power coupe, which was produced from 1917 to 1918. It had a four-cylinder gasoline engine next to an electric motor. Woods had been manufacturing electric cars since 1899 and the company attempted the hybrid to stay in business. But the car was expensive and its fuel economy was not an advantage since gasoline was not expensive. Few were sold.

Before the electric car died the speed and range had been improved. The last Detroit Electrics had a competitive top speed of 35 miles per hour by the early 1920s. The light Dey runabout of 1917 was available at just $985.

By 1926, there were more than 8 million automobiles in America along with a new coast-to-coast federal highway (U.S. Route 40). By then the electric car had lost most of its support.

Early Fords

Henry Ford started up his first working gasoline engine in his kitchen in 1893. But by the time he set up the Ford Motor Company up in 1903, Oldsmobile, Cadillac, and Buick were established businesses. Ford was born on a Dearborn farm. He had a keen interest in mechanical things, set out to build cars that were simple to operate and affordable. The Ford Model T went from $950 in 1909 to $240 in 1925. Ford's farming background helped him to understand his customers' needs and by 1918 almost half of the cars in the world were Model Ts. The Ford assembly lines were supplying 10,000 cars per day.

When Henry Ford began to use assembly lines in auto production to reduce costs in 1914, he was able to pay some of his employees five dollars per day, which was a relatively high wage at the time. By 1929, almost half of all U.S. families owned an automobile, this percentage was not reached in England until the late 1960s.

There was some opposition to the automobile, the first vehicles were noisy and scared horses. Some Minnesota farmers even plowed up roads but most Americans greeted the car with tremendous enthusiasm. The automobile quickly replaced the horse and wagon as a more efficient means of transportation.

As consumerism grew it reorganized America's relationship with farming and the land in a way that transformed the car from a riding luxury to a goods gathering necessity. Much of one's food and clothing was once produced in the home, but by the 1920s it was bought in towns and villages. By the 1930s, 2/3 of rural families and 9/10 of urban families purchased store-bought bread instead of making it on their own. More shopping resulted in more time spent in cars and people's priorities changed.

The car provided consumers with a convenient means of going to the market and stimulated the growth of suburbs which began as far back as the 1840s with the start of railroad travel. This growth received a jolt after the Civil War with the advancement of streetcar lines.

Late in the nineteenth century, major cities such as New York, Chicago, and Philadelphia became factory centers. After World War I the construction of office buildings in the cities instead of new housing, pushed many middle-class residents from the city to the suburbs, using cars to take them to work. By 1940, almost 15 million Americans lived in communities not serviced by public transportation.

The automobile met the need for transportation and cars freed people from train and streetcar schedules, allowing them to travel at their own pace. Automobiles were marketed with this thought of freedom and immunity in mind. Advertising often shows someone in the comfort of an automobile free from the worry of snow or rain. Independence is the message being conveyed.

Some recognized the problems of the automobile as the smell of unburned gasoline grew as early as 1905. Gasoline caused noticeable pollution and its status as a nonrenewable resource was questioned. Engineers and industry analysts began to wonder if an adequate supply would remain available with the growing popularity of the automobile.

Alternative fuels such as grain alcohol were available. But, alcohol was more expensive at twice the price per gallon at the turn of the century. This did not include the federal excise tax placed on alcohol in 1862 to help reduce the Union's costs in the Civil War. By 1907, the tax was repealed. But, the process of denaturing alcohol, to make it undrinkable and enforce the sobriety of Americans, added to its price and gave gasoline the advantage. It also took more alcohol to produce the same amount of power than gas.

The Car Industry Grows

The United States had 8,000 registered cars and trucks in 1900 and almost 200 million in 1995. This growth started with the efforts of hundreds of companies, but it become almost the exclusive domain of Ford, General Motors and Chrysler in the United States and a handful of companies in Europe and the rest of the world.

As the industry grew up in the U.S., there was a tremendous entrepreneurial evolution in the area around Detroit and other parts of the country following the turn of century, as the expectations of private automobiles and unending roads grew.

Many of the early cars, were first built in bicycle and wagon shops. In Detroit, almost 140 auto companies were formed from 1900 to 1903, but about half of these would fail by 1904. The founder of General Motors, William Durant, began as a successful wagon maker. GM began when a young inventor named David Buick, assured Durant that horseless carriages would replace horse-drawn wagons. Buick was building bathtubs and liked to work with engines.

In 1904, the first year of operation, the Buick Motor Company sold 37 cars. Durant was involved with Buick after 3 months of its startup. The early years were difficult but 1907 Buick was the second largest carmaker in the country. Durant foresaw the demand for cars when many others did not. He incorporated General Motors in 1908 and bought Oldsmobile, Cadillac and Oakland by 1909. The company acquired part suppliers Champion Spark Plug, Delco and Fisher Body. In 1910, it controlled almost a fourth of the U.S. auto industry. Durant introduced auto air conditioning and started the General Motors Acceptance Corporation to allow cars to be bought on credit.

Henry Ford was the father of mass production, but it was Alfred P Sloan, Jr. the president of General Motors, who introduced the annual model change as a way to keep the consumer in a new car with features the older vehicles did not have. Under Sloan, GM passed Ford as the nation's number one auto producer. The company succeeded not by offering consumers a basic means of transportation (Ford's plan) but by offering faster cars that grew with more style and power with every year. It resulted in GM's dominance in the industry for decades.

LEADED GAS

Leaded gasoline was a key factor in engine performance. Early cars had to be cranked by hand to start. But in 1911, the invention of the self-starter eliminated the need of hand cranking. Automakers could now produce larger cars with larger, easy-to-start engines. As larger cars were produced, customers noticed a knocking sound when the engine was stressed climbing hills or accelerating. If cars were going to be larger, faster, and easier to use, then a way had to be found to eliminate potential engine damage from engine knock.

Shortly after the invention of the self-starter, engineers at Dayton Engineering Laboratories Company (DELCO) found that ethanol or grain alcohol could be used to reduce knock. The problem with grain alcohol was that anyone could make it. In 1921, Thomas Midgley was working at the DELCO lab, now owned by GM, when he found that tetraethyl lead was an excellent antiknock compound.

By 1923, leaded gas was being pumped at Dayton, Ohio. By the following year, GM, DuPont Chemical Company (which controlled about a third of GM's stock) and Standard Oil of New Jersey, combined their various patents and produced leaded gasoline under the Ethyl brand name.

A few months before Ethyl went on sale, William Clark of the U.S. Public Health Service stated tetraethyl lead was exceedingly poisonous and had the potential to produce enough lead oxide to endanger public health in heavily traveled areas.

In 1923, General Motors, financed a study by the U.S. Bureau of Mines on the safety of tetraethyl lead. The bureau issued a report downplaying leaded gasoline's potential adverse impact on public health.

With the burning of large quantities of gasoline starting in the 1950s, lead deposits were appearing in soil. The widespread use of leaded gasoline was presenting a health hazard. Other more agreeable additives existed in ethyl alcohol blends. Leaded gasoline allowed Detroit no boost performance and sell automobiles, but even radioactive waste breaks down over time.

In the United States, seven million tons of lead were released between the 1920s and 1986, when it was phased out as automakers switched to unleaded gasoline and catalytic converters.

PUBLIC TRANSPORT

The auto's dominance over mass transit did not result in the end of public transportation. Some cities such as St. Louis, New York, and Chicago showed an increase in riders during the '20s. GM's effort to urge consumption through model changes and faster, more stylish cars was partly a response to the continued strength of public transportation.

Streetcar companies throughout the nation were undermined with municipal regulations that forced them to keep fares low and often prevented the elimination of unprofitable routes. Unionized transit workers compelled the companies to pay high wages. GM spent 18 months in the mid-1930s putting New York City's trolley system, one of the world's largest, out of business, by substituting buses.

In 1936, GM formed National City Lines, a group made up of Firestone Tire and Rubber, Phillips Petroleum, Standard Oil of California, and

Mack Manufacturing, the truck company. Over the next decade, the group took control of almost 40 transit companies located in 14 states. It also acquired a controlling interest in several other companies located in four additional states. In 1947 a grand jury indicted the company for violating the Sherman Antitrust Act (1890).

Franklin Roosevelt's New Deal was spending enormous amounts of money into building roads but little on mass transit. Starting as early as 1916, the Federal Road Act made funds available to states to establish highway departments. Legislation passed in 1921 established the Bureau of Public Roads and planned a network of highways linking cities of more than 50,000 people. But, it was during the New Deal that major road-building began.

Almost half of the two million people employed in New Deal programs worked in constructing roads and highways. During the 1930s, the total amount of surfaced roads doubled, to more than 1.3 million miles, while mass transportation languished. Public transit received only a tenth of the money that the Works Progress Administration spent on roads.

America was becoming an asphalt nation due to weaknesses in the streetcar industry, the push to sell buses, and the state-sponsored building of roads during the Depression.

ENERGY TRENDS

The suburban home with its car, lawn and undetached house uses nonrenewable fossil fuels. In the 25 years following World War II, American homes consumed increasing amounts of energy. In the 1960s, energy use per house rose by 30%.

But the high-energy home was not inevitable. Even in the 1940s many popular magazines featured articles on solar building design. Innovative homes were oriented toward the south to use the sun's heat in the winter and had large overhangs to reduce the effects of the summer sun. They saved natural resources and appealed to America's wartime conservation attitude. Even in the late 1940s and early 1950s, solar homes received serious attention from architects, builders, and the media. The World War II era represented a time of resource conservation. But, as the 1950s bloomed, the availability of cheap heating fuels like oil and natural gas reduced the need to conserve. The federal government reduced investments in solar design and research efforts and the use of coal, oil, and gas for heating increased.

Suburban home energy use also increased as many discovered air conditioning. In 1945, few American homes had air conditioning, even though the technology was available since the 1930s. As air conditioning units became cheaper and more compact in the late 1940s, sales boomed.

Once the wartime housing shortage ended in the mid-1950s, builders used air conditioning to increase the demand for new homes. The addition of air conditioning in new homes allowed buyers to trade up. Air-conditioned offices helped fuel the demand for central air conditioning in homes.

The National Weather Bureau, in 1959 started issuing a Discomfort Index, providing a measure of heat and humidity. The air conditioning industry grew as the index became more popular.

Between 1960 and 1970, the number of air-conditioned houses grew from one million to nearly eight million. These energy using units added certain comforts to the home and made life more tolerable in many areas of the country. It also helped to reduce the effects of suffering from heart or respiratory diseases.

The growth of suburban home building also had other effects on energy resources. As large-scale builders cleared the land of trees, the new homes were subject to more heat and more cold, increasing the energy that was required to keep the temperature comfortable inside.

Suburbia turned into a sea of green as homeowners bought lawn-mowers, pesticides, fertilizers, and sprinklers to maintain a lush green carpet. Oil and natural gas are the chief components in the production of nitrogen-based fertilizer. Power mowers use fossil fuels and contribute to air pollution. An hour of mowing grass can produce as much pollution as driving a car 350 miles.

AUTO EFFICIENCY

Although smaller cars use less petroleum, they are not necessarily more efficient. Of the energy released in combustion only 12-15% is finally applied to move the car. Most of the rest is lost due to the thermodynamic inefficiency of the engine and escapes as heat. The remainder is drained off by such factors as aerodynamic drag, rolling resistance of tires, transmission slippage, internal friction, idling, and air conditioning.

Just to push the air out of its way, a car may use 50% of the available

energy at 55 mph and 70% at 70 mph. Large frontal areas create air turbulence and drag. Bodies derived from wind tunnel testing can provide
a more smooth air flow around the vehicle. Details such as mirrors, rain
gutters, trim, wheel wells and covers can be more appropriate.

Radial tires can reduce fuel consumption as much as 3%. Punctureproof tires of plastic could save even more and eliminate the cost and
weight of a spare tire and wheel.

Automatic transmissions inflict a mileage penalty of about 10% compared to manual gearboxes. Continuously variable transmissions can provide better mileage. A stop-start engine that shuts down if a car is idling or
coasting can cut gas consumption by about 15%. A touch on the accelerator restarts the engine.

Better lubricants and bearings can reduce friction, and microprocessors can monitor systems and make adjustments to keep operation at peak
efficiency without or despite actions by the driver.

The proper maintenance of roads can improve mileage 5%. Since
combustion engines operate best at about 45 MPH, traffic ideally should
be speeded up in cities and slowed down in the country. This will prove
to be difficult though possible, low highway speeds are unpopular and
higher city speeds are impractical for safety reasons.

As the price of petroleum for gasoline and diesel engines converges
with that of alternate energy sources, new power systems will become
widely available. Battery-powered electric motors have their advantages;
quietness, low pollution and simplicity. But, their disadvantages of limited range between recharges (which are also limited), weight, and bulk
reduce their market potential. New battery systems could give better performance but they have not been forthcoming. Efficiency can also be increased by using flywheels to equalize power demands on batteries during acceleration and hill climbing.

Electric motors are paired with small combustion engines in hybrid
systems with electric power for low speeds and combustion for highway
passing. Hybrid systems offer much better mileage and have proved to be
popular beyond the estimates of most auto manufacturers.

Power systems that run on compressed gases such as propane, methane, or hydrogen are problematical. Range may be limited since distribution systems are not in place and each station pump could cost $30,000.

So-called synthetic fuels could be used directly in engines or to generate electricity from fuel cells for electric motors. Other combustion engines such as the sterling motor may also become options.

Electric Revivals

Electrics have made some reappearances over the years. One was after the fuel problems from the Arab oil embargo in 1973. The utility-endorsed Henney Kilowatt of 1959 to 1961 was a converted French Renault Dauphine. About 120 Henneys were built. Other efforts were not that successful.

1970 Era Electrics

Out of the 1973 oil embargo came a number of strange looking electric cars. These included the Free-Way Electric, which looked like a bug, the 3-wheeled Kesling Yare, with styling straight out of Starwars and the B&Z Electric, which seemed to have been made of wood scrapes.

Florida-based Sebring-Vanguard sold 2,200 of their 2-seat, plastic body CitiCar following the 1973 oil embargo. They resembled a phone booth on wheels and were powered by golf-cart batteries with a top speed of 30 miles per hour and a range of 40 miles in warm weather. Viable commuter EVs would appear with the conversions of the late 1980s.

Steam Revivals

There have also been revivals of the steam car. Robert McCulloch, the chain-saw millionaire, spent part of his fortune on a steam prototype, called the Paxton Phoenix, between 1951 and 1954.

William Lear of Learjet fame, spent $15 million in 1969 on a turbine bus and a 250-horsepower turbine steam car. Both used quiet, efficient steam engines although the bus had reliability problems and poor gas mileage. Lear also tried to enter a steam car into the 1969 Indianapolis 500. The British firm of Austin-Healey was also working on a steam car in 1969. It had four-wheel drive. However, even wealthy entrepreneurs like McCulloch and Lear found that they lacked the means and support structure to successfully mass market a competitive car. Alternative power systems would have to wait until air-quality regulations resulted in some breakthroughs with hybrid and even fuel-cell cars.

During the 1950s, the Volkswagen Beetle became a popular small car and by 1960 had sales of hundreds of thousands. Other economy cars included compacts such as the Ford Falcon and Dodge Dart. In Europe, 50-miles-per-gallon microcars like the BMW Isetta and Morris Minor enjoyed some success.

Most modern electric cars in the 1960 to 1980 era were subject to failure, with pop-riveted bodies and non-existent marketing. The Urba Sports Trimuter of 1981 was 3-wheeler with a pop-up canopy, needle nose and a top speed of 60 miles per hour that the owner could build from a set of $15 plans.

Other creations included the Free-Way Electric, which looked like a bug, the 3-wheeled Kesling Yare, with styling out of A ClockWork Orange and the B&Z Electric seems to have been made of wood scraps.

Florida-based Sebring-Vanguard sold over 2,000 of their 2-seat, plastic body CitiCar following the 1973 oil embargo. They resembled a phone booth sized box on wheels and were powered by golf-cart batteries with a top speed of 30 miles per hour with a range of 40 miles in warm weather. Viable commuter EVs would have to wait until the conversions of the late 1980s.

Air Conditioning

The addition of air conditioning in new homes allowed buyers to trade up. Air-conditioned offices helped fuel the demand for central air conditioning in homes. The National Weather Bureau, in 1959 started issuing a Discomfort Index, providing a measure of heat and humidity. The air conditioning industry grew as the index became more popular.

Between 1960 and 1970, the number of air-conditioned houses grew from one million to nearly eight million. These energy using units added certain comforts to the home and made life more tolerable in many areas of the country. It also helped to reduce the effects of suffering from heart or respiratory diseases.

The growth of suburban home building also had other effects on energy resources. As large-scale builders cleared the land of trees, the new homes were subject to more heat and more cold, increasing the energy that was required to keep the temperature comfortable inside.

Suburbia turned into a sea of green as homeowners bought lawn-mowers, pesticides, fertilizers, and sprinklers to maintain a lush green carpet. Oil and natural gas are the chief components in the production of nitrogen-based fertilizer. Power mowers use fossil fuels and contribute to air pollution. An hour of mowing grass can produce as much pollution as driving a car 350 miles.

ENERGY GROWTH

The spreading of economic development to all the reaches of the globe has fueled the growth of the automobile. Most industrialized nations including Japan, Britain, Germany, France and others have seen great changes in energy growth as well. But by the end of the 20th century, the United States used more energy per capita than any other nation in the world, twice the rate of Sweden and almost three times that of Japan or Italy. By 1988, the United States, with 5% of the earth's population, consumed 25% of all the world's oil and released about a quarter of the world's atmospheric carbon.

CARBON ACCOUNTING

Almost 4.5 billion years ago, the earth was formed, and 95% of the atmosphere consisted of carbon dioxide. The appearance of plant life changed the atmosphere since plants, through the process of photosynthesis, absorb carbon dioxide. Carbon from the atmosphere was drawn into the vegetation. When the vegetable matter died, it decomposed, and formed coal and oil. This reduced the carbon dioxide in the atmosphere to less than 1%.

Industrialization and the burning of fossil fuels reverses this process. Instead of being drawn out of the air, carbon is extracted from the ground and sent into the atmosphere. In the United States, a major surge in energy consumption occurred between the late 1930s and the 1970s, rising by 350%. More oil and natural gas was used to meet industrial, agricultural, transportation and housing needs. Oil and natural gas contain less carbon than coal or wood, but the demand for electricity and fuel soared as the nation's economy grew and consumers became more affluent.

By 1950, Americans drove three-quarters of all the world's automobiles and they lived increasingly in energy consuming suburban homes, with inefficient heating and cooling systems. Appliances were also inefficient. A 1970s era color television operated for four hours a day was the energy equivalent of a week's worth of work for a team of horses. U.S. energy consumption slowed down in the 1970s and 1980s, as manufacturers introduced more efficient appliances. Even so, by the late 1980s, Americans consumed more petroleum than Germany, Japan, France, Italy, Canada and the United Kingdom combined.

EARLY CAR DESIGN

The manufacturing process in early cars resulted in their conservative appearance. The chassis were produced by the new automobile companies, but the bodies were built by tradition-oriented carriage-builders. The technology of the craftsmen in these firms was wood-based and they built car bodies, like carriages, on a wooden frame. Metalsmiths, painters and trimmers completed the process. The level of elaboration and surface decoration was linked to the social status of the customer at this early stage.

In this age of the evolving car the carriage-builders created a box to fit on the chassis. It housed the car's components and provide a place to sit. Gradually, changes occurred to the car's structure and the auto industry. Improved manufacturing processes forced the carriage trade to fade away and the age of the automotive engineer began. Aluminium and then steel replaced wood. Soon the all-metal car appeared in 1914 as Dodge introduced an all-metal model, which was quickly emulated elsewhere. Across the Atlantic, the Italian maker Lancia built an all-steel car body in 1918 and the French firm Citroen marketed an all-steel saloon in 1928.

The use of steel for structural members and body panels facilitated the move toward prefabrication, standardization, and mass production. In the American automotive industry this change occurred in the teens of the century led by Ford. The carriage trade followed this lead to some extent, developing standardized car bodies which were supplied to the large chassis producers.

Even more significant was the shift from a chassis-based automobile to a unit-built car. Developments in metal technology made this possible, allowing cars to be single structures rather than chassis/body combinations. Vincenzo Lancia launched his unitary Aurelia model in 1933, followed by the Aprilia. The British firms Vauxhail and Morris and the German companies Opel and Adler produced their own unitary cars. Production methods changed to provide this development, it was another move father away from carriage building.

The unitary car started a revolution in car design. For the first time a car could be conceived as a single entity. In 1904 the Panhard Levassor Company arranged the car's key components and set the architecture of the modern car. Developments such as the raising of the hood and the lowering of the sitting position of the passengers provided the car with its strong horizontal moving emphasis. Advances in racing cars responded to the emerging area of aerodynamics and resulted in the use of the torpedo body.

By 1914 the modern motor car had become almost completely distinct from the horseless carriage. Car designers embraced a visual simplicity and applied it to the car. The idea of the machine aesthetic moved into the world of the automobile and produced the ultimate consumer machine. Other forms of transport, such as aircraft and boats also evolved into modern forms to facilitate their movement through air and water. This design movement treated cars for their status rather than machines. Cars became objects of inspiration.

The impact of aerodynamic theory was important. In 1914 Paul Jaray was working for the Zeppelin airship company and began to evolve his ideas about aerodynamic which were to have a huge impact on car design decades later.

The automobile developed a modern visual identity of its own beyond one of visual simplification and technological breakthroughs. Its role as a key market of social status in the first half of the 20th century meant that there was an emotional side to its visual impact. In France, high-class carriage-builders pushed the form of the car to new levels of fantasy in the inter-war years, extending bonnets and developing long sweeping curves linking the fenders to the running boards. Complex mouldings, multiple colors, and luxurious interior fittings reinforced the idea of the car as a fashion accessory. From then on, the push of technology and the pull of fantasy and desire worked alongside each other in modern cars, making the automobile an icon and a symbol of indulgence and consumer decadence.

Mass production brought the car within the reach of many people in the years following World War I. Henry Ford's Model T brought the car to the mass market but was only available in black. General Motors took the lead in the late 1920s, showing that mass-produced cars could be objects of desire.

The late 1920s represented a turning point in modern car design.

When General Motors hired custom-car designer Harley Earl to head its new Art and Color Section, this prototype in-house styling studio started an approach to designing cars that was to be emulated worldwide and would dominate the industry for the rest of the century. Earl was an important part of this, but his work coincided with several developments that made a significant difference. One was the Dupont company's new cellulose paint, called Duco, which could be sprayed on car bodies, eliminating extensive drying and finishing processes.

The late 1920s was a time when manufacturers became more aware of women in car purchasing. This had a major effect in the design of cars. Women were thought to value aesthetic and comfort properties over performance. As a result, visual design aspects took on a pronounced significance for the large manufacturers.

In the late 1920s and early 1930s many popular cars combined smooth curved surfaces from aerodynamic studies with the integration of components into the main body of the car. The visual influences ranged from aircraft and boats to racing cars. By the start of the World War II years, the coachbuilders had disappeared, except in Italy.

Earl and his group at General Motors established many of the practices that were adopted. In recent years the use of computers has dramatically modified the methods evolved by Earl and his successors.

Car designers have been targets of those that see the automobile as an enemy of the environment and of modern life. However, designers were seeking to satisfy a market which existed. The result was this close relationship between the car and society. Many designers today attempt to address the problems of pollution and diminishing resources. Along with other designers of the 20th century, car designers played a role in making technology culturally acceptable.

Building cars involves large teams of people each bringing a different set of skills to the task. The design of cars is subject to technological, social, cultural, economic, ergonomic and political forces.

THE EARLY FORD YEARS

A major step toward the age of modern mass-produced automobiles occurred at the end of the first decade of the 20th century. Henry Ford abandoned the workshop practices in use and adopted the moving assembly line. This breakthrough transformed the manufacturing process.

In 1913 Ford put in place the final piece in a solution that had been emerging in the last years of the previous century. Evolving techniques from flour mills, arms manufacturing and the meat-packing industry combined into the use of standardized, interchangeable components and unskilled labor for large-scale manufacturing. It replaced the skilled craftsmen who produced the highly finished, handmade early automobiles.

The machines that resulted off Ford's production line at his Highland Park factory in Dearborn were cheap, functional vehicles directed at the huge working population of the United States. Those living in the rural areas would be especially transformed by this new ease of mobility.

Ford developed his ideas for his first automobile in 1883. In 1896 he built his first car and in 1903 he started the Ford Motor Company. In 1908 his Model T appealed to the masses for its functional simplicity, hard-wearing qualities and ease of maintenance. It was not a beautiful car and its appearance dictated a combination of component parts rather than a unified silhouette but its utilitarian features gave it a high level of appeal.

The Age of Utility

Ford understood utility and the appeal of his Tin Lizzie was in its reliability. Customers could have any color as long as it was black. This promised the owners of his car that every Model T was as good as every other Model T.

Ford became a leading manufacturer in the teens of the 20th century. He paid his workers well and was able to lower the price of his cars, from $850 in 1908 to $260 by 1925. By that year two million Models Ts had been sold and he built a new factory at River Rouge.

By 1927 his production had risen to 15 million. By the late 1920s, the unchanging Model T became a issue. General Motors had a newer approach to styling which made the Model T look old-fashioned.

Ford closed his River Rouge plant for a few months and developed his more stylish Model A, but his lead over his competitors was never again as great. From the late 1920s on Ford cars would be restyled on a regular basis like General Motors and Chrysler vehicles.

THE ERA OF HARLEY EARL

In 1927, Harley Earl started the Art and Color Section at General Motors Company. Henry Ford had the vision to realize the potential for car ownership and in the early part of the century ownership of a Model T Ford has enough to confer a high level of status on these customers. Harly Earl went beyond this, he realized that once car ownership was more universal, Americans would want more than reliability and low price. He transformed the automobile from an engineered object to a stylish artifact. This was a dramatic conceptual shift that has never been overturned. He went beyond the rational needs of consumers to a deeper desire and turned the car from a utility object into a realizable dream. Harley Earl had a talent for visualizing dramatic car bodies. He created a car for Fatty Arbuckle and one for Tom Mix that even came with a saddle.

Affordable Luxury

Alfred P. Sloan Jr. became president of General Motors in 1923 and saw a huge expansion of the company through that decade so that it overtook Ford as the leading automobile manufacturer in the United States. It achieved this by focusing on a newer way of selling cars to the American market. Instead of providing a cheap, standardized model it promoted product diversity, annual model changes, installment purchasing and trade-in. Also, Sloan offered a car for every purse and every purpose. The ideas of style and comfort had been available only in custom-made, luxury vehicles. Sloan wished to offer every person a luxury car at a price they could afford. The new urban consumers could differentiate themselves from others.

The luxury automobile of the inter-war years was still designed in the coachbuilding tradition. Thus, the new Chrysler Airflow failed to capture much of the market.

Until the 1950s, the wealthy customers required traditional values in their cars to demonstrate their social status. This trend reached its peak in the mid-1930s, with the classic cars of that era.

Early in the 20th century, luxury car manufacturers existed in most European countries as well as in the United States. The early luxury cars in Britain included Rolls-Royce, Bentley, Daimler, Jaguar, Lagonda and Aston Martin. In France there was Bugatti, Delage and Delahaye. Germany had Horch, Maybach and Mercedes while Italy had Isotta Fraschini and Spain Hispano Suiza. In the United States there was Cadillac, Packard, Duesenberg, Pierce Arrow and Lincoln.

Some of these were grand tourers and while others gradually transferred their focus from the racing scene to consumers. All of these sought to convey an image of luxury and refinement that was forward-looking. As standardized production geared up in increasing numbers, there was more need at the top end of the market, to stress individuality and distinction. The manufacturers relied on providing every possible luxury.

By the 1920s high style was reflected in Harley Earl's LaSalle of 1927. It was influenced by the Hispano Suiza. The Depression made the market tighter, but these styles lasted through World War II.

Then, cars appeared more modern in appearance, with unified and sculptural body forms. There was more use of chrome as surface decoration and the influence of streamlining was noticeable. Although many cars maintained some dignity from the past and had conventional features such as straight hoods and gently sloping grilles and windshields.

A technology-led transformation took place in car manufacturing and the large car manufacturers found cheaper ways of making cars. The one-piece body and chassis meant the end of traditional coachbuilding techniques.

The post-World War II years featured cheaper, mass-produced cars which seemed to incorporate the same features as earlier luxury cars. Europe was also turning its car manufacturing into a mass-production industry. Fiat in Italy, Citroen and Renault in France, and Morris and Austin in England introduced more efficient production methods in their factories.

The push for a standardized automobile that everyone could afford was not just an economic goal. In Europe it was also political. The creation of a car for the people was a symbol of the masses promoted by fascist and liberal governments. These people's cars included the British Morris Eight and the Italian Fiat Topolino.

In the United States Ford and Chrysler tried to emulate the luxury cars of Cadillac and Packard. But, the difference between the automobiles aimed at the rich and the poor in Europe remained more distinct. Luxurious, chauffeur-driven cars were produced by Rolls-Royce and Hispano Suiza and contrasted with tiny cars produced for the low end of the market.

Cars for the masses tended to be as small as possible. By 1945 the VW Beetle, the Fiat 500, the Citroen CV, the Morris Minor and the Renault 4CV had emerged. Low fuel consumption was also a key requirement of these vehicles. During these years many workers in Europe exchanged their bicycles and motorcycles for cars.

The postwar European people's car also became a mark of national identity. The small cars of the 1950s had some features in common but they were all distinctive and created for the roads that had to carry them, such as the efficient autobahns of Germany or less developed rural roads of France. Few cars moved across national borders in these years. A number of low-priced bubble cars appeared but many of these were less than stable.

BUBBLE CARS

The 1950s saw the introduction of a number of very small, three-wheeled cars that were known as bubble cars due to their rounded forms and curved-glass windows. Their attraction was low price and low fuel consumption. Also, their owners did not have to pay a high car tax. They were easy to park and represented a step up from a motorcycle and side-car. Parents could get a child into some of these vehicles and they were used as a second car for some families.

The little Messerschmitt 175 was produced by the former German aircraft manufacturer. It had a Plexiglas canopy which was lifted for entry. Later versions used a pivoting front seat for better entry and exit. The Messerschmitt had two wheels in front and one in the rear. It could accommodate an adult and a child. More features were added later including soft-top and four-wheeled versions.

The Italian Isetta had two front wheels and one version with a single back wheel and another with two back wheels close to each other. BMW took over production of the Isetta in 1954, a year after its introduction in Italy, and produced it until 1964. By then 160,000 cars had been made. Entry was through a large front door. The steering wheel was designed to pivot when the door was opened. Versions included one with a bubble rear window and another with a sliding side window. A convertible was also produced. In 1957 BMW introduced its four-wheeled 600 model. The 600 had a door in the front and another to allow entry to the back seat. Through the 1950s many small companies entered the field. Britain had the Petite, Clipper, Frisky, and Scootacar.

There was a hazard in driving these unstable three-wheelers, but a number of Japanese microcars appeared in the 1990s as the need to lower fuel consumption made them appealing once again.

In the early 1950s, the production of several mini and three-wheeler cars included Suzuki's Suzulite of 1955 and Fuji's Subaru 360. The Japanese people's car was a product of the late 1950s and 1960s. Toyota's Publica model of 1961 represented this direction as well as Nissan's Datsun Bluebird.

By the 1980s Japan had become the world's second largest manufacturer of automobiles, after the United States. Larger cars such as the Nissan President and the Toyota Century had existed since the 1960s. Japan would target its cars at different lifestyles based on a deep understanding of the changing market for cars.

When World War II ended, Willy-Overland decided to produce the Jeep for the civilian market, starting with the production of its CJ (Civilian Jeep) line in 1945. The first Jeep station wagon was the Wagoneer in 1946. It was the first all-metal station-wagon. A Jeep truck appeared in 1947, and a convertible, the Jeepster in 1948.

The Jeepster combined the front end of the Willys station wagon with the rear fenders of the Jeep truck. Its flat-topped fenders became a key characteristic of later versions. Less than 20,000 had been manufactured when it went out of production in 1950.

The Jeep and its variations was a part of the American scene from the mid-1940s. It retained its original identity as a utility vehicle for the consumer market. In 1974 the Cherokee appeared as the first of the sports utility vehicles (SUVs), Kaiser took over Willys-Overland in 1953 and Willys wagons and trucks continued to be produced until 1965. The company was taken over by American Motors in 1970 and then by Chrysler in 1986.

Another utility vehicle that reached international status was the British-designed Land Rover. Maurice Wilks was the chief designer at Rover at the end of World War II. He was using an ex-army Jeep on his estate and would need a replacement when the car wore out. He had several Jeeps taken apart at the Rover factory and then built a four-wheel-drive vehicle for the British market. Since there still were restrictions on steel, except for the export market, the body of the Land Rover was made of aluminium alloy. The prototype was built on a Jeep chassis in 1947 and the final model appeared in 1948. It was a success with British farmers and gradually moved into the general consumer market.

The original Land Rover (Series 1) was a basic vehicle with no thought to physical comforts. The headlights were behind a mesh grille with the sidelights on a bulkhead below the windshield. Like the Jeep, the body-shell was made of flat panels. The Rover company was also producing several sedans that were successful in the British marketplace, including the Rover 75.

In 1970, Rover combined the utility of the Land Rover in rough terrain the comfort and luxury of its sedans. The result was the Range Rover, which was produced until 1995. The Range Rover proved to be a success as a rural car which fit into the urban lifestyle. It was comfortable and performed well in rough terrain.

Both the American Jeep and the British Land Rover were highly successful with consumers. They symbolized a working society that used material goods as tools rather than as display icons. While car styling was becoming more extravagant, this was an alternative that had a strong appeal for many. In rural and urban areas, these utility vehicles could serve several purposes. But, by the end of the century this concept proved to be largely symbolic.

This type of vehicle was widely emulated in the 1980s and 1990s. The concept of the sports utility vehicle (SUV) grew quickly and was endorsed by manufacturers around the world. The Jeep and the Land Rover saw major gains in the consumer market. In the late 1980s Chrysler converted the Jeep into a definite urban car. Its flat fenders were replaced by a more rounded style. The Land Rover was transformed into the Discovery, Defender and Freelander for the new urban markets.

Japanese manufacturers picked up the SUV concept and the Toyota Land Cruiser, Mitsubishi Shogun, Isuzu Trooper, Mitsubishi Pajero, and Mitsubishi Chariot Grandis appeared. These high-off-the-ground, four-wheel-drive, large-wheeled vehicles were highly successful.

THE ROLE OF THE SUBURBS

The auto was a major factor in the expansion of the suburbs in the 1950s. America had a corresponding need for increased mobility. A car was needed to do the shopping and increased affluence made the purchase of a car possible for many who had never owned one before.

For the first time, the market became fragmented, with husbands, wives and teenagers targeted separately by the auto manufacturers. Car advertising became rampart in the press and on television.

The late 1950s and early 1960s saw a new desire for power and speed. New kinds of automobiles appeared—compacts, personal luxury cars, and pony cars. These competed with the sedans that had dominated in an earlier time.

The American automobile culture was attacked in the late 1950s by Vance Packard and in 1965 by Ralph Nader in a book called *Unsafe at any Speed*. The growing safety concerns led to the passing of the National Traffic and Motor Vehicle Safety Act in 1968.

The oil crisis and the rising environmental movement of the early 1970s applied the brakes to the age of gas-guzzlers. To some the car ceased to express the good life and began to become a threat to the future of society.

The cars of the late 1930s and early 1940s were large, bulky objects with headlights integrated into massive wings, large chrome radiator grilles and bumpers that curved around the bodies. After War II, the imagery moved towards items of military technology and transportation like the fighter jet and rockets.

The cars in America from the late 1940s to the early 1960s were reproduced on an annual basis, like fashion goods, in their attempts to outdo each other. They featured incremental improvements which led to bigger and better cars for the consumer.

THE AGE OF FINS

This competition was noticeable in the application of fins to the rear ends of cars. This trend was started by Harley Earl who was given the task of differentiating the Cadillac from the other General Motors lines. Earl added some small fins on the rear of the car.

These fins introduced a new look in American car design. The LeSabre of 1951 was named after the F-86 Sabre fighter jet. The Lockheed P-38 Lightning fighter plane had three fuselages and aerodynamic tailfins. Earl introduced this dramatic look into the Cadillac Sedanet of 1948. Other manufacturers followed and by the mid-1950s tailfins had become familiar in many American cars. The curved windshield was inspired by the cockpits of planes and several cars of the late 1940s had bullet noses that were suggestive of aircraft.

New colors such as pink, pale green and lemon yellow appeared to seduce consumers. Labor-saving devices, such as power steering and electric windows also became popular. Lush interiors were directed at female consumers but the jet fighter imagery of the exterior was masculine.

By the mid-1950s, General Motors had introduced fins on all their cars. The next half decade saw fins rise to even greater heights. The Chrysler imperial had exaggerated tailfins in 1955. Chrysler claimed that the fins had aerodynamic properties that improved performance.

When Earl brought out his Cadillac Eldorado in 1959, it had the largest fins so far with rocket tail lights. The Eldorado Brougham was equipped with everything a passenger could desire, including a built-in powder compact and drink containers.

The 1950s were dominated by dreams that money could buy. The postwar economic boom brought the automobile within the reach of more people than ever before. This new generation craved luxury and needed to impress and outdo their neighbors. By the mid-1960s, car designers would try to rethink the image of the private car and construct a different market.

MUSCLE CARS

The mid-1950s saw the appearance of American sports cars, such as the Corvette and the Thunderbird. The car market was fragmented with station wagons, sedans and sporty cars.

From the 1930s onward many young men opted out of mainstream car buying, constructing and racing their own hot rods or custom cars. This sense of needing something special began to affect the mass market and led to the emergence of muscle cars and pony cars. They lasted for less than a decade, but the muscle cars were powerful and fast and literally had visual muscles in the form of curves and bulges on their body surfaces.

The first muscle car was the Pontiac GTO. Others include the Firebird, the Trans Am and the Chevrolet Camaro. The Ford Mustang of 1964 was the first pony car. The basic model was simple and had no frills, but it was sporty and could be enhanced with a wide range of accessories. It was inexpensive and became a big market success.

Other muscle and pony cars followed, including the Ford Fairlane GT, Dodge Charger, American Motors' AMX, Buick Skylark GS and the Chrysler Barracuda.

By the early 1970s muscle and pony cars had faded. They had become bigger and heavier and started to look like each other.

RECENT TRENDS

The fears of car safety and pollution in the 1960s were followed by the sudden oil crisis of the early 1970s. The anti-car lobby saw the car as a symbol of modern life to be attacked rather than adored.

The 1950s had been a time when optimism prevailed and the exuberance of that era had pushed the design of automobiles beyond aesthetic limits. The effect of the oil crisis produced a move towards economy and utility, as manufacturers strove to make their cars as functional as possible. High style was succeeded by aero aesthetics primarily for reasons of fuel efficiency.

A higher level of seriousness and caution came to dominate the approach to cars. Manufacturers found themselves working in a field of more regulations and safety related data.

The financial difficulties of the 1980s had the effect of creating a designer culture in domestic goods. Consumers believed that they needed to create and absorb themselves into their own individual lifestyles. This movement was seen in the automobiles of the 1990s.

Japan had much to do with the new range of automobiles from the multi-purpose vehicles (MPVs) to the sports utility vehicles (SUVs) and their variations. Different market sectors desired different types of cars. The increasing number of women and young people driving cars was a factor in this mode of thinking.

The ideas generated by Japan spread quickly to the United States and Europe. The car was now a lifestyle accessory. Technologists developed the use of new materials, new safety systems and new forms of in-car communications.

Brand identity would express the essence of the vehicle, the way it sat on the road, how it made the driver feel and the nature of its appearance or its demeanor.

As the car manufacturing industry became increasingly global, luxury brands moved in new more dynamic directions. In the United States macho cars, rooted in the heritage of drag racing and car-customizing saw a resurgence while Japan, with its congested cities, developed new mini-cars. Popular classics such as the Volkswagen Beetle and the BMC Mini were revived.

In the 1980s and 1990s, companies such as Nissan, Toyota, and Mazda created highly original cars such as the Nissan Figaro, the Nissan Pao, and the Mazda Maita. Microcars from Japan and Korea in the 1990s, included the Suzuki Wagon, Daihatsu Move and Daewoo Matiz.

In Japan, several of the leading manufacturers became established in the years between the two world wars. Mitsubishi was producing cars in the 1920s while Nissan and Toyota built cars in the 1930s. The Toyota model AA of 1936 was modeled after the Chrysler De Soto Airflow.

During the first decade of the 20th century in the United States and Europe, motoring was still an upper-class leisure activity and most cars were chauffeur-driven. The availability of the electric car in the first decade of the 20th century gave women their own automobile for use when shopping or visiting. Charles Kettering's invention of the automatic self-starter was known as the ladies aid.

World War I brought many women into closer contact with the world of transportation and the number of female drivers increased through the inter-war years but not as fast as the number of men. There was a shift from utilitarian and economic values to an emphasis on beauty, comfort and social status.

The fuel crisis of the 1970s started a movement in car design towards a more utilitarian vehicle. In the 1990s, there was a movement to the character car. The technological aspects were now so widespread that customers could no longer make purchasing decisions on this basis alone. Also, it had become practically impossible to distinguish between different models. Manufacturers perceived that a new emotionalism was entering the world of car purchasing and that consumers were being led by their hearts rather than their heads. Symbolic meanings were acquiring a new level of significance.

Eco-Cars

As the number of cars continued to grow, one of the challenges for car manufacturers was to respond to the regulatory requirements related to auto safety and pollution. More than any other single product the car was responsible for problems in air pollution through exhaust emissions and the use of gasoline as fuel. These problems had been growing since the mid-1960s. Alternative fuel options include the reemergence of the electric car and the use of hydrogen, either in liquid or fuel cell form.

BMW has made progress with liquid hydrogen and has manufactured several models in its new 7 series that can run on this fuel. It is stored in a tank behind the rear seats. Ford has teamed up with DaimlerChrysler, and General Motors with Toyota, to develop cars with hydrogen fuel cells. See Table 3-1 for a summary of major auto manufacturers.

Table 3-1. Major Automobile Manufacturers

Audi AG
Ingolstadt, Germany
A 99%-owned subsidiary of Volkswagen AG. It has a range of cars; the A3, A4, A6, A8, and the TT sports car.

Automobili Lamborghini Holding SpA
Bolognese, Italy
A subsidiary of Volkswagen AG. Less than 300 vehicles made a year.

Bayerische Motoren Werke AG (BMW)
Munich, Germany
Automobiles make up 75% of BMW's sales, also produces motorcycles.

Daewoo Motor Company Ltd.
Inchon, South Korea
Part of the family-run Daewoo Group, Korea's second largest automobile producer behind Hyundai, makes cars, buses, and trucks.

Dahatsu Motor Company Ltd.
Osada, Japan
Known for its small, inexpensive cars.

Daimler Chrysler AG
Stuttgart, Germany
Result of the merger of Mercedes-Benz and Chrysler in 1998, cars manufactured include Dodge, Eagle, Jeep and Plymouth.

Ferrari Spa
Modena, Italy
Sports cars, such as the Testarossa.

Fiat SpA
Turin, Italy
Small cars such as the Cinquecento, also owns Alfa Romero, Ferrari and Maserati.

Ford Motor Company
Dearborn, Michigan
America's largest manufacturer of pickup trucks, besides Ford, Mercury and Lincoln also owns Aston Martin, Volvo, and part of Mazda.

General Motors Corporation
Detroit, Michigan
Largest producer of cars and trucks, owns Buick, Cadillac, Chevrolet, GMC, Pontiac, Saab, Saturn, and Oldsmobile.

Honda Motor Company LTD
Tokyo, Japan
Japan's third biggest manufacturer, makes motorcycles as well as cars.

Hyundai Motor Company
Seoul, South Korea
Produces cars, minivans and trucks, Korea's leading producer of vehicles.

Isuzu Motors LTS
Tokyo, Japan
Known for producing pickup trucks and engines for boats and tractors.

Jaguar LTD
Coventry, UK
Producer of luxury sports vehicles

Mazda Motor Corporation
Hiroshima, Japan
Fifth largest Japanese manufacturer, makes cars, minivans, trucks and Miata sports cars.

MG Rover Group LTD
Birmingham, UK
Makes SUVs.

Mitsubishi Motors Corporation
Tokyo, Japan
Japan's fourth largest car manufacturer, produces a range of passenger cars including the Galant, Aspire and Eclipse. Also produces trucks.

Nissan Motor Company LTD
Tokyo, Japan
Second largest Japanese manufacturer, produces passenger cars, pickups and SUVs. Renault owns 37% of Nissan.

PSA Peugeot Citroen SA
Paris, France
France's largest vehicle manufacturer and second largest in Europe, produces cars and light commercial vehicles.

Pininfarina SpA
Turin, Italy
Also creates products such as telephones and watches.

Dr. Ing, HCF Porsche AG
Stuttgart, Germany
Produces the Boxster and 911 models, also manufactures watches and luggage.

Renault SA
Boulogne-Billancourt Cedex, France
France's second largest manufacturer, 44% is owned by the French government.

Saab Automobile AB
Trollhattan, Sweden
A subsidiary of General Motors, produces a range of cars, including the 9.3 Turbo and the 9.5.

Seat SA
Barcelona, Spain
Started in 1950 by the Spanish government and Fiat, now a subsidiary of Volkswagen.

Suzuki Motor Corporation
Shizuoka, Japan
Japan's leading producer of minicars.

Toyota Motor Corporation
Aichi, Japan
Japan's leading vehicle manufacturer and fourth largest in the world.

Vauxhall Motors Ltd.
Bedfordshire, UK
Owned by GM

Volkswagen AG
Wolfsburg, Germany
Europe's leading car manufacturer, owns Audi, Lamborghini, Rolls-Royce, Bentley, Seat and Skoda.

AB Volvo
Gothenburg, Sweden
Sold its car operations to Ford, produces trucks and buses.

Electric cars are less attractive to the mass market. The worry of having to plug them in to recharge them is combined with the dread of coming to a sudden halt in an inconvenient place when the batteries run out of power.

Hybrids

The hybrid car combines a gasoline engine with an electric motor. Honda launched its hybrid Insight model in 2000. It combined a 1960s-style futurism which is reinforced by the innovativeness of its technology. A gasoline engine in the front is combined with an electric engine in the rear. Since the batteries add to the weight of the vehicle, weight reduction was accomplished with a body made of aluminium and nylon. The Insight was the first car to offer the hybrid solution to the general public. It advantages are the 30 km/liter (83 mpg) fuel economy and the generation of about half of the carbon dioxide of comparable small cars.

Toyota also produced a hybrid car, the Prius, and has been working on a hybrid minivan and SUV. The problem is in developing a car with the appropriate technology that also appeals to consumers. The challenge is to create an aesthetic that is of the moment and in keeping with the spirit of an age in which people value the planet and its resources.

In the 21st century, new initiatives are visible in car design. Many possibilities present themselves, some rooted in past developments, others exploiting new possibilities that link technological, social and cultural issues. Many depend on new technologies; the use of fuels other than gasoline, computer-based power control and braking systems as well as navigation systems for enhanced safety.

ALTERNATIVE FUEL TECHNOLOGIES

DaimlerChrysler's electric NeCar 4 has been widely promoted but yet cannot counteract the disadvantages of electric power. DaimlerChrysler is also working on hybrid cars. Opel General Motors has built a hydrogen-powered car, HydroGen 1. Ford plans to make its SUVs lighter and more aerodynamic and had plans to introduce an electric version of its Escape SUV. VW's Lupo is designed for low fuel consumption. Pininfarina's Metrocubo of 1999 is a hybrid car.

Car safety may depend on technology providing many of the answers. Mazda has proposed an ASV (advanced safety vehicle), which has a monitor screen to show rear and side views, voice-interactive navigation system and uses collision avoidance technology.

In the area of new computer technologies, Ford has its 24.7 concept car with a computer console in the dashboard. The 24.7 allows complete Internet access which is voice-activated.

Other technological innovations include the move to new materials to provide fuel-efficiency through lightness. Aluminium and plastics will be used more extensively.

Nissan's Hypermini is made almost entirely of these materials. These materials are also combined in new ways. The Dutch company Hoogovens has worked with I.D.E.A. SpA to build a car from a laminated sandwich of aluminum and plastic. Among the new car designs in Japan was the Toyota Will-vi concept, a minivan nostalgic of Citroen's 2CV. A similar lifestyle concept car is the Honda Fuya-jo which also appeared in 1999. This is a tall vehicle with semi-standing seats.

In the United States cars have been getting bigger and more aggressive. In Japan they are getting smaller, more boxlike and taller. One development is an emphasis on interior space with more room and more flexibility combined with a basic exterior. Here the exterior becomes more of a container than a symbol of power and speed.

In the search for new markets manufacturers have been creating new kinds of cars by merging existing model types. The SUV has merged with the MPV and the MPV with the microcar. As society becomes more complex and niche markets more specialized, manufacturers seek new formats to meet the needs of lifestyle shifts. There is an awareness that cars play an important role in the formation of personal or group identity. Several recent concept cars are structured in new ways with doors opening in a novel manner and interiors with a high level of flexibility.

The American Automobile Manufacturers Association which merged into the international Alliance of Automobile Manufacturers claims that today's automobiles are up to 96% less polluting than cars 35 years ago but automobiles still produce a quarter of the carbon dioxide generated annually in the United States.

A global accord on reducing hydrocarbon emissions was reached at the 1992 Early Summit in Brazil. Only Great Britain and Germany have come even close to meeting their 2000 targets. The United States was short of its goal by 15 to 20%. How could this admirable effort at international cooperation succeed, when cars are the major nonstationary culprits and almost every country is filling its roads with more and more of them?

The international agreement on global warming signed by 150 countries in Kyoto, Japan, late in 1997 requires a drastic reduction in automobile exhaust emissions. The world's largest producer of carbon dioxide emissions is the United States. Greenhouse gases were to be reduced to 5.2% below 1990 levels by 2012. There has been much opposition to Senate ratification of the Kyoto accords. The objectors say Kyoto is based on questionable science and would damage the U.S. economy. It exempts two of the world's biggest polluters, China and India.

Kyoto must be backed by 55 countries representing 55% of the world's emissions to go into effect. The question is not if the greenhouse effect exists, it pivots on the theory that emissions have an effect on global warming.

Meeting the Kyoto goals would not be easy, and the auto industry would have to do its part. Tightening of the corporate average fuel economy (CAFE), the federal standard for cars and trucks would be needed.

A 12 mile per gallon car or truck emits four times as much carbon dioxide as a 50 mile per gallon subcompact. The auto industry has fought against attempts to tighten CAFE and until 1998, when California cracked down, has kept sport utility vehicles exempt from regulations.

Actual fuel economy has declined since 1988, as the car manufacturers switched from producing more fuel-efficient smaller vehicles to more profitable but larger trucks and sport utility vehicles (SUVs).

The Coalition for Vehicle Choice (CVC), which is a lobbying group sponsored by carmakers, has campaigned to repeal the CAFE standards. The CVC has claimed that CAFE causes 2,000 deaths and 20,000 injuries every year by forcing people into smaller cars.

The oil and auto industries have questioned the science behind global warming causes and claimed there was not enough information. That may be changing, Toyota was the first auto company to announce, in 1998, that it was joining others such as British Petroleum, United Technologies, and Lockheed Martin in an alliance to battle global warming. Toyota is also supporting the Pew Center on Global Climate Change, which was started with a $5 million grant from the Pew Charitable Trusts.

References

Sparke, Penny, *A Century of Car Design*, Barron's Educational Series, Inc.: Hauppauge, NY, 2002.

Chapter 4

Fuels for the Auto

In 1900, when there were only a few thousand motor vehicles registered in the United States, the public could choose from steam, electric, or gasoline automobiles. Although we take it for granted today, a gasoline-based transportation system was by no means a foregone conclusion then. To a public used to horses, the idea of sitting on top of a boiler, battery or flammable gasoline tank, making forward progress by a series of dramatic explosions, was not instantly appealing.

At least 50 years in the future, pollution was not an issue. None of the early advertisements for electric cars touted clean air benefits, although the fact is that they ran silently and did not soil their operators' clothes.

More than a century later, the auto industry is at a similar crossroads, with competing technologies. Just as in 1900, there will be turns in the road, technical breakthroughs and factors that no one has thought of. New propulsion systems will emerge and the internal-combustion engine may become a historical artifact, after what may turn out to be, in history's long sweep, a relatively brief hundred year run.

The newer, more efficient, cleaner cars will emerge not from a wagon maker's garage, but from well-funded research projects. The new technologies range from the relatively familiar hybrids with both gas and electric power to the more exotic fuel cell vehicles which are, in essence, electric cars. The future presents unique challenges and opportunities on the way to successful commercial application.

The newer gasoline cars may be considered as high-tech wonders, but they are still rooted in the 19th century. As modern materials, electronics, and manufacturing processes improve, they make the vehicles of today closer to the theoretical ideal of the old technology.

AUTO TECHNOLOGY

It is not easy to start a car company and it gets more difficult as auto technology advances. Before the Depression, there were automakers like

McDonald, McFarlan, Mercer, Meteor, Mitchell, Monitor, Monroe, Moon and Moore and these were just the M's.

The huge drop in personal income during the Depression years killed hundreds of small manufacturers. Most lacked enough capital to build and market complex products. After World War II there was much pent-up demand and owning a car became a reality for most families, but the independent makers found it hard to compete.

Hudson survived the war years and brought out its advanced streamlined models in 1948. But, it could not update its model quickly enough to compete with Ford, Chrysler and General Motors. The 1957 Hornet marked the end of the Hudson line.

Even Hudson was huge compared to the electric vehicle start-ups that appeared after the 1973 Arab oil embargo. Companies like Linear Alpha produced conversions along with Electric Fuel Propulsion, which sold its 100-mile-range Transformer model for a short time. Sebring-Vanguard made its CitiCar and was able to sell all it could make for a short time, but sales slumped when gas prices dropped and the gas lines disappeared.

One of the most affordable of the electric cars was the Danish-made Kewet El-Jet I. It sold at Green Motorworks in North Hollywood, California for $18,350 fully loaded. This was at a time when most EVs started at $25,000. But, instead of a Ford Taurus body, there was an offensive fiberglass box that looked like an enclosed golf cart.

Driving the Kewet was reported as similar to the Yugoslavian Yugo or the East German Trabant. Performance was slow and noisy and the car had little attention to detail. Old style sliding windows were used and while most electrics are very quiet on the road, the Kewet was loud with road noise and rattles. There was a large front window which steamed up during heavy rains since the defroster was not effective.

As electric vehicles became available, do-it-yourself projects became less common, but there were many. John Stockberger of Chicago converted an old Pinto using a surplus military aircraft generator and a burned-out shell found in a wrecking yard. In California Bill Palmer used golf-cart batteries to power an old Chevy for an 80-mile range. Clarence Eller built his Aztec 7 XE, which was a futuristic looking sports car with gull-wing doors. It served as an advertisement for his electric vehicle conversion manual.

Ron Kaylor, Jr., was an electrical engineer from Menlo Park, CA, who started building electric cars in the early 1960s. He specialized in VW Beetle conversions using motors from F-100 fighter planes. In the early 1970s, he offered his Kaylor Hybrid Module that provided VW electric cars with a 400-mile range.

Hybrid conversions have not been as common, since they are twice as complex as electric conversions. But in 1979, Dave Arthurs of Springdale, Arkansas, spent $1,500 converting an Opel GT into a hybrid that got 75 miles-per-gallon, using a 6-horsepower lawnmower engine, a 400-amp electric motor and a bank of 6-volt batteries. The editors of *Mother Earth News* also built a hybrid, using a 1973 Subaru. It achieved almost 84 miles per gallon. This was immediately after the 1973 Arab oil embargo. When *Mother Earth News* offered the plans, there were 60,000 requests for them. Dave Arthurs continued building hybrids into the 1990s. One of these conversions was a 99 miles-per-gallon Toyota pickup which used a 9-horsepower diesel engine.

In California one of the larger electric car builders was U.S. Electricar in the 1970s. It converted the Renault LeCar to electric power and by 1994, the company was publicly traded, with 300 workers in three plants of 170,000 square feet. Its Los Angeles factory was converting Geo Prizms and Chevy S-10 pickups, but in 1995 the company had large losses and complaints of defective cars. It seems that the company expanded rapidly without a solid base of sales and it disappeared quickly.

The Solectria Company started on a small-scale in 1989. Its start was in selling solar panels, electric motor controllers, and converters to college electric vehicle racing teams. This led to electric conversions on compact cars and pickup trucks. Even the most basic conversion was $33,000 and most sales were to utilities and government agencies.

Solectria had government contracts from the Defense Advanced Research Projects Agency (DARPA) and won a bid to build its lightweight Sunrise EV. The production of 20,000 Sunrise cars a year seemed like a possibility. This kind of volume would have made Solectria the major electric car producer, but it was not likely to achieve it without help from a major automaker partner. But, just as Solectria was looking for a partner, the auto companies' own electric programs were being launched. Although the Sunrise was considered to be an inventive design, the Detroit automakers had their own designs in cars like the GM EV1.

The Sunrise did not make it into production and only a few prototypes were built. Solectria did go on to build the Force EV, which is a converted Chevrolet Metro and was involved in the GM CitiVan, which was an urban delivery vehicle.

Other independent electric vehicle companies included Unique

Mobility, which became known for former Mouseketeer Darlen Fay Gillespie's manipulation of the company's stock, and conversion companies like Green Motorworks and the Solar Car Corporation. Some of these companies had ties to retired Detroit automakers. Former Chrysler Chairman Lee Iacocca raised $10 million to start EV Global, a company that built electric bikes and other vehicles. EV Global bought batteries from Energy Conversion Devices, a company where former GM Chairman Robert Stempel served as a consultant and public spokesman.

One of the best designs came from independent Paul MacCready who used his design breakthroughs in lightweight race cars in electric vehicles at his California R&D firm, AeroVironment.

Many firms were hopeful that they could compete with Detroit, but Solectria had to price its Force EV based on a fully equipped Chevrolet Metro that the company had to buy at retail from a local dealer. It took Solectria seven years of building this car before it was able to buy engineless cars from GM.

By the late 1990s, it was clear that only the big automakers could make the electric car really happen. But, these were the same companies that had scorned electric cars earlier. GM sued California in the U.S. District Court in Fresno to block imposition of the state's zero-emissions rules. These regulations would require automakers to build thousands of electric vehicles using rechargeable storage battery technology. But, the auto industry contends that conventional, electric-powered cars are too expensive and too limited in range to be profitable.

MARKETING CHALLENGES

Of the 300 million cars in the United States, only a few thousand were highway capable electric vehicles and some of these were conversions. Most dealerships do not put much effort in marketing alternative fuel vehicles because of the limited demand. Battery electric cars suffered from their limited range and recharging stations. They were only marketed in a few states with very limited advertising. Even with this limited effort, sales were disappointing. Honda discontinued its EV Plus program and by the spring of 1999, after three years on the market, GM leased only 650 EV1s and 500 S-10 electric pickups. Toyota's RAV4 EV was on the market since the end of 1997 and only sold 500 vehicles by 1999. Ford sold about 450 Ranger electric pickups in the same time period and only about 250 leases were signed for the EV Plus. In 10 years an EV maker, Solectria sold only 350 converted cars and trucks. Honda viewed the EV Plus as a successful data exercise since it provided data from a wide customer base.

Many companies choose to lease vehicles to commercial fleets to limit their risk from limited sales. Many including California's CALSTART consortium thought the lease numbers would be higher if there were more cars available.

Most opinion polls show that the motorist's infatuation with automobiles does not include internal-combustion engines. Many drivers would trade in their current car for an electric vehicle, if it could perform as well and not cost any more. One poll of California new car buyers conducted by the University of California at Davis in 1995 found that almost half would buy an electric vehicle over a gasoline car, but they wanted a 300-mile range and a more reasonable price.

Most commuters have round trips of 50 miles or less, but a longer distance is important for trips and visits. Accessories such as air conditioners, power windows and locks tend to limit an electric car's power and range even more.

Cost is always a problem when vehicles are made in limited numbers since the parts will cost more. The lithium ion batteries used in Nissan's Altra EV were reported to cost close to six figures. Since electric cars sell for $30,000 or more, a lease can soften the cost of the vehicle. It also isolates the user from expensive battery replacements. Even these subsidized leases required an extra $100 or more in monthly payments compared to a more conventional vehicle. Leasing allows the manufacturers to keep control of the vehicle for repairs and recalls. As the technology changes, a lease keeps customers from having a 2-3 year vehicle that is out of warranty with needing obsolescent, expensive parts.

The Ecostar van was Ford's first electric since the time of Thomas Edison and Henry Ford. The Ecostar was the first electric vehicle that resembled an actual production car instead of a conversion. It had a recharge port and a battery charging meter.

The Ecostar provided a pleasant driving experience comparable to a quiet luxury car. An electric powered vehicle can be extremely quiet and will appear to be as transparent to drivers as possible.

The Ecostar used high-temperature sodium-sulfur batteries because of their range, but they also allowed the van to go from 0-60 in 12 seconds. The Ecostar had no trouble keeping up with gas vehicles, but Ford only built about 80 Ecostars. The sodium-sulfur batteries were too sensitive to cold weather. They operated at 500° Fahrenheit and started fires in several of the demonstrators.

U.S. Electricar built the lead-acid Electricar Prizm in Torrance, California at Hughes Power Control Systems. This GM subsidiary also designed the car's DC-to-AC inverter. Instead of a gas gauge there was a range meter. The batteries were in a covered tunnel underneath the car.

Most electric vehicles have good low-end torque for excellent 0-60 acceleration, but the Prizm was a little sluggish initially but then picked up quickly. The car used a recharging paddle.

Detroit's electric cars have a shaky history since the market can change rapidly in the automobile industry, which is dependent on long lead times for new models. In 1975, when memories of the oil embargo were fresh, Detroit's cars were still growing in size, but it was a record sales year for the Volkswagen Beetle and sales of Toyotas and Hondas sales reached 100,000 that year. GM's profits dropped 35% and the company had to temporarily close 15 of its 22 assembly plants.

With nothing but full-sized cars in its inventory, GM launched a crash program to build an economy model, which resulted in the Chevette. It was based on GM's German Opel Kadett, a 4-cylinder, 52-horsepower compact with 35 miles per gallon economy. In 1976, GM sold almost 190,000 of the hatchback Chevette.

At GM, the market seemed right for electric vehicles. Small start-up companies had been offering electric conversions for commuter vehicles. The CitiCar was produced by Sebring-Vanguard, which for a short time was the fifth largest automaker in the United States.

GM would build an EV called the Electrovette in 1980. It replaced the Chevette's 4-cylinder engine with a DC electric motor and zinc nickel oxide batteries. The Electrovette used a mechanical controller. The batteries were expensive and not much better than lead-acid power for extending the range of operation. The Electrovette had controller problems and GM let the project die.

GM would launch the EV1 in 1996 and show a number of alternative-fueled concept cars at the 1998 Detroit Show. The EV1 assembly line was in the old Buick Reatta plant, next to the much larger and more automated facility that assembled Chevrolet Cavalier and Pontiac Sunfire convertibles. Some 30 employees essentially hand-built the cars on the line and traded off tasks.

GM built efficiency into their electric cars. They worked on reducing energy consumption, mass, and accessory loads, and improving aerodynamics, rolling resistance, and driveline efficiency. There is a 50-kilowatt fast charger, which can give an EV1 a charge in 10 minutes. An aluminum space frame is used which allows the body, without batteries, to weigh in at 1800 pounds.

Almost every part of the EV1 is designed for energy efficiency. The steering wheel and seat frames are made of low-weight magnesium. The radio antenna is part of the roof to reduce drag. The tires are low-rolling-resistance, self-sealing Michelins which also saves the weight of the spare. The aerodynamic body sits only five inches off the ground. There are 2000 spot welds in the aluminum body. The cars came with air conditioning and CD players standard and were sold through Saturn dealers.

The 1300-pound battery pack sat on a 1500 pound body resulting in a total weight of about 2800 pounds. Getting the batteries to produce more power, weigh less, and take up less space was one of the goals in cars like this.

These same concepts could be used in fuel cell powered cars. Ultralight fuel cell vehicles are a part of the current generation of clean concept cars, sometimes called Green Cars.

GM made several things happen when it marketed the EV1. The 1998 California zero-emission mandate was one result of GM stating that it would have an electric car on the market by the mid-1990s.

In 1996, when EV1 became available, it accelerated the development of the hybrid Toyota Prius. Now every auto show has its alternative cars.

FUEL CELL ELECTRIC VEHICLES

The fuel cell was first demonstrated in principle in 1839 but there was no practical application during that time period. Sir Robert Grove was involved in patent cases and would often suggest improvements in a products' design. Grove had a laboratory, where he made several important improvements to the design of storage batteries. He also invented the fuel cell and described how the chemical combination of hydrogen and oxygen could be used to produce electricity.

A fuel cell car, bus or truck is in essence an electric vehicle powered by a stack of cells that operates like a refuelable battery. A battery stores chemical energy, but a fuel cell uses an electrochemical process to generate electricity and operates from the hydrogen fuel and oxygen that are supplied to it. Like the plates in a battery, the fuel cell uses an anode and cathode, attached to these are wires for the flow of current. These two electrodes are thin and porous.

Most automotive fuel cells use a thin, fluorocarbon-based polymer to separate the electrodes. This is the proton exchange membrane (PEM) that gives this type of fuel cell its name. The polymer provides the electrolyte for charge transport as well as a physical barrier to block the mixing of hydrogen and oxygen. An electric current is produced as electrons are stripped from hydrogen atoms at catalysis sites on the membrane surface. The charge carriers are hydrogen ions or protons and they move through the membrane to combine with oxygen and an electron to form water which is the main by-product. Trace amounts of other elements may be found in this water, depending on the cell construction. In most cells the water is very pure and fit for human consumption. Individual cells are assembled into modules that are called stacks.

PEM fuel cells can convert about 55% of the fuel energy put into them into actual work. The comparable efficiency for IC engines is in the range of 30%. PEM cells also offer relatively low temperature operation at 80°C. The materials used make them reasonably safe with low maintenance requirements.

The emergence of commercial fuel cell cars will depend on developments in membrane technology, which are about one third of the fuel cell cost. Improvements are desired in fuel crossover from one side of a membrane to the other, the chemical and mechanical stability of the membrane, undesirable side reactions, contamination from fuel impurities and overall costs.

One recent breakthrough in membrane technology occurred when PolyFuel, in Mountain View, CA, produced a hydrocarbon polymer membrane with improved performance and lower costs than the current perfluorinated membranes. This cellophane like film has performed better than more common perfluorinated membranes, such as Dupont's Nafion material. The hydrocarbon membrane can also operate at higher temperatures, of up to 95°C, which allows the use of smaller radiators to dissipate heat. It also lasts 50% longer, while generating up to 15% more power and operating at lower humidity levels.

Fluorocarbon membranes can cost about $300 per square meter, the PolyFuel materials should cost about half of this. While hydrocarbon membranes have yet to prove themselves to many, Honda's latest FCX fuel cell cars use them.

Catalysts

Another key part of a PEM membrane is the thin layer of platinum-based catalyst coating that is used. It makes up about 40% of the fuel cell cost. The catalyst prepares hydrogen from the fuel and oxygen from the air for an oxidation reaction. This allows the molecules to split and ionize while releasing or accepting protons and electrons.

On the hydrogen side of the membrane, a hydrogen molecule with two hydrogen atoms will attach itself to two adjacent catalyst sites. This frees positive hydrogen ions (protons) to travel across the membrane.

The reaction on the oxygen side occurs when a hydrogen ion and an electron combine with oxygen to produce water. If this is not controlled properly, highly corrosive by-products such as hydrogen peroxide can result, which quickly damage the internal components.

In a proton exchange membrane (PEM) fuel cell, protons travel through a film 18 microns thick. This is the proton exchange membrane. Electrons are blocked by the film and take another path which provides the electric current flow. Over time and usage tiny holes can form on the film which reduces fuel cell performance. If you strengthen the film, then you also reduce performance.

Ballard Power Systems had one of the first fuel cell demonstration projects with British Columbia Transit. Ballard is a major producer of fuel cells, and it has installed them in city buses. The fuel cell powered New Flier buses are much cleaner than new diesels and they are not contributing to Vancouver's smog problem at all. A similar demonstration has been under way in Chicago, where the modified city buses have been given the name Green Machines.

The Vancouver pilot program was the world's first real test of fuel cell vehicles. The buses are quiet except for the whirl of their air compressors and have a range of 250 miles.

Ballard does not build cars, trucks, or buses. Its sole product is the fuel cell in all of its many applications, plus the ancillary equipment to make it work. Because Ballard's technology is so advanced, its fuel-cell sales make it one of the fastest-growing automotive suppliers in the world. Ballard's alliance partners include DaimlerChrysler, Ford, Honda, Nissan, Mazda, Volvo, and Volkswagen.

The company was founded in 1979 to build rechargeable lithium batteries for smoke detectors. It was started by a Canadian engineering geologist. Geoffrey Ballard also worked for the U.S. Department of energy. In 1983, Ballard was approached by the Canadian Department of

Defense who were interested in fuel cells. Since they were similar in operation to batteries, they thought Ballard might be interested in a developmental contract. There are now research operations in Germany and California as well as several facilities in Canada. Before Ballard actually made a profit (except under Canadian accounting rules), its stock made meteoric gains.

Ballard should have a fuel cell ready for volume production, up to 250,000 annually. Many questions involve fuel cell availability and much is dependent on the auto industry. Fuel cells are starting to appear in autos, even with the limited infrastructure available now.

In Ballard's alliance with Ford, Volvo, and DaimlerChrysler, they will supply the other components of the vehicles from the car body to the electric motor drive. The fuel cell will function as the car's engine. It needs cooling, control and fuel processing.

Fuel Cell Future

The ultimate goal is a fuel cell car that is competitive in price and performance with the internal combustion vehicle. Some early users will pay a premium for new technology. But most drivers will not pay 20-30% more for similar performance. If fuel cell and internal combustion cars have the same refueling, power, and convenience features, and one costs much more than the other, it will suffer.

There may be cost-effective fuel cell prototypes, but what of the infrastructure to support it? A main problem is fuel storage, and some are doubtful that the high-density storage of hydrogen gas will be practical very soon. Liquid hydrogen also has problems.

Methanol would allow a transitional phase where some fuel cell vehicles use methanol, which is relatively simple to reform and would not present too big a change from our current system. However, methanol is toxic and very corrosive. Some gas stations would need to be retrofitted to operate with it (fuel tanks and fuel lines replaced.) But, many gas station tanks are already methanol-compliant.

Fuel cells are overdue to become a major player in our energy future. Then, one in three or even one in two of the cars on the road may be fuel cell vehicles. There are obstacles and challenges to that happening, but there does not seem anything insurmountable. The cost of materials is being reduced and high-volume manufacturing will bring production costs down.

A hydrogen-based economy could be the ideal scenario for personal transportation. Hydrogen has a lot of advantages but one problem is the hydrogen infrastructure. Hydrogen is difficult to handle since it is flammable at 4% concentrations.

A decreased reliance on fossil fuels, with electrics, fuel cell cars, hybrids and much-improved internal-combustion engines is possible. The internal combustion engine may not remain dominant, but it not clear what technology will replace it. The mass production of fuel cell cars is some time away. If cost-competitive fuel cell stacks are available soon, it will change the competitive mix.

The present period can be likened to 1900, when gasoline, electric, and steam cars all competed for market share, with the public and the industry unsure of the future. Fifty years from now the fossil fuel era may be seen as a distant memory just like we look at the steam age now.

Fuel Cell Advances

Delta Airlines is using a hydrogen fueled tow tractor at the Orlando Airport. General Motors has delivered its first fuel cell truck to the U.S. Army. The U.S. Navy plans to use fuel cells for ship-board power with hydrogen sourced from diesel fuel. New fuel cell technology from SANYO may power IBM Thinkpad notebook computers for 8 hours.

Dow Chemical and General Motors are installing up to 400 fuel cells at Dow plants. Hydrogen is a natural by-product at Dow and will provide 35 megawatts at its facilities.

John Deere is testing fuel cell modules for off-road applications. Hydrogen is considered a good replacement for diesel in locomotives. Recent testing indicates that it could be economical for railroads.

GM has been using the EV1 as the base for its next generation of hybrid and fuel cell cars, while Ford is working on the P2000 lightweight sedan as a development vehicle. It uses Ford's aluminum 1.2 liter direct-injection DIATA engine and achieves 63 miles per gallon. The hybrid model has an even higher mileage. Ford has been at work on a car powered by direct hydrogen. It was equipped with a 5,000-psi compressed hydrogen tank, but this would only provide a range of 50 miles, although the acceleration was excellent. Ford will be reducing even more weight off the car along with other improvements. The P2000 was one of the world's few operational fuel cell cars when it was completed.

DaimlerChrysler's ESK2 debuted at the Detroit Auto Show in 1998. It was a lightweight, aerodynamic 70-miles-per-gallon hybrid car and provided a platform for alternative fueled vehicles including fuel cells.

Reforming
DaimlerChrysler has been an advocate of gasoline reforming. Using available fuels will allow fuel cells on the market more quickly. Hydrogen could be processed from gasoline onboard vehicles until hydrogen becomes a more practical fuel choice. Politics may decide what fuel is used. Ethanol, for example, can be made from corn and is popular in the Midwest.

DaimlerChrysler is predicting that an onboard sensor would tell what kind of fuel is being pumped in and then adjust the reformer on the fly. This system would have some complexity, since different fuels are reformed at different temperatures, using varying proportions of steam and air. Other companies are not taking this approach, although most agree that reformers should be adaptable for the variety of fuels used internationally. Brazil would need ethanol reformers and the U.S. methanol or gasoline models. Most countries have one common fuel.

An onboard hydrogen tank has several problems. Hydrogen leaks easily, is hard to store and hard to compress and burns quickly. Overcoming all these concerns may be costly. Refueling may tend to be difficult although there are a number of hydrogen refueling stations in use around the world.

The HyNor project in Norway has plans to build a hydrogen highway between Oslo and Stavanger on the southern coast of the country with refueling stations spaced along the route. Iceland has plans to start with a small fleet of fuel cell buses in the capital, Reykjavikk, then slowly convert every vehicle on the island even fishing boats creating the world's first hydrogen economy.

REINVENTING THE AUTO

GM, along with others, has been working at reinventing the auto. GM developed its AUTOnomy and Hy-wire concept cars. Now, with the fuel cell Sequel, GM has been able to double the range and half the 0-60 mph acceleration of these cars in less than three years.

The Sequel is almost the size of a Cadillac SRX. It has a 300-mile range on a refueling of hydrogen and accelerates to 60 mph in less then 10 seconds. Other fuel cell cars have a driving range of 170-250 miles and cover 0-60 mph in 12-16 seconds depending on whether they use a battery.

All of the drive power of the Sequel is in an 11-inch-high chassis. The individual powered wheels provide excellent control on snow, mud, ice and uneven terrain. GM's start-up time in freezing conditions is less than 15 seconds at -20°C. GM has working systems now and knew that if they are going to put these cars into the marketplace, they would have to start in the middle of a northern winter.

GM believes that it could eventually close down engine and transmission factories around the world and have a single plant making fuel cells for all of its vehicles. There are 29 types of engines made in 28 GM plants worldwide and 20 transmissions made in 20 worldwide plants.

A fuel cell vehicle requires only 1/10 the parts needed for internal combustion models. A change to fuel cell power could end overcapacity problems for GM. It would no longer have to consider different state or country environmental regulations. Fuel cells also free designers and allow them to be more creative with styles and body designs.

GM has pledged to develop a hydrogen fuel cell vehicle that would compete with conventional cars in volume by 2010. The company has 1,000 people working on the project in government, university and private labs in 14 countries. It has spent over $1 billion on the project since 1996. DaimlerChrysler has also spent a billion on hydrogen fueled technology.

GM's international Global Alternative Propulsion Center is responsible for developing fuel cells for world markets. The center has several operations in Germany, where in concert with German subsidiary Opel, it has built the Zafira fuel cell minivan.

The Ford Ecostar van program was launched in 1993 in response to the announcement of GM's EV1. Ford also started building fuel cell prototypes, but they were not really road-ready vehicles.

In 1997, Ford announced that it would invest $420 million in a global alliance with what was then Daimler-Benz and Ballard Power Systems. This provided Ballard with an important infusion of capital. As a result of these investments, Ford owned 15% of Ballard and DaimlerChrysler 20%.

It was a critical moment for fuel cells since the total investment was reaching almost $1 billion, including the $450 million by DaimlerChrysler. The alliance of Ford, Volvo, and DaimlerChrysler was pushing the leading edge of fuel cell innovation. Ballard has focused on PEM cells with a goal to have commercial fuel cells available by 2010.

In Germany, Daimler-Benz's fuel cell prototypes included the NE-CAR III. This was a Mercedes-Benz A-Class car with Ballard's methanol-reformed fuel cell system.

DaimlerChrysler's newest fuel cell is in the Mercedes-Benz b-class car. The fuel cell is a sandwich design with the polymer PEM cell between two gas permeable electrodes of graphite paper. Hydrogen is introduced to one side of the fuel cell while the other side is exposed to the air. Like the GM Hy-Wire platform, the fuel cells, fuel tank and fuel systems are under the floor. The compressor is in the front of the car to reduce the noise. There are four hydrogen sensors on the fuel cell stack, on each of the hydrogen tanks, another at the electric motor and another inside the cabin.

The high torque electric motor develops more than 100-kW of motive power which is 35-kW more than the previous design for the A-class. The fuel cell is also more efficient. An enhanced hydrogen storage system gives the vehicle a range of 250 miles (400-kM). The Ballard fuel cells are expected to last at least 5,000 hours in a car and 10,000 hours in a bus.

Ford had plans to go into production with a fuel cell family car based on the aluminum and composite P2000 which is like the Ford Contour but weighs a thousand pounds less. In 1997, Ford announced that its fuel cell car would carry compressed hydrogen, but the fuel storage question may be still open.

One of Ford's partners is the Virginia-based Directed Technologies consultant firm. They have advised Ford that cars can carry hydrogen gas, eliminating the need for costly and bulky reformers. Along with onboard hydrogen storage, they also believe that the problems of building the hydrogen infrastructure can be overcome.

Studies point to the superiority of direct hydrogen, but this is regarded by some in the industry as less attractive than liquid fuels such as methanol. A large steam reformer plant could supply 1 million cars with hydrogen.

If methanol is used directly, there has to be an onboard reformer and a revised infrastructure to deliver it. But methanol does have some advantages. There is excess generating capacity, and it's the least expensive fuel to transport.

Some 70% of the world's oil is in OPEC countries, and 65% of it is in the Persian Gulf. If we switch to methanol, which is produced from natural gas, we can diminish that dependency.

A truly zero-emissions hydrogen generating system using solar or natural sources is popular where the fuel is produced from an aggregate of photovoltaic collectors, wind generators, and biomass. This would allow a motor vehicle fuel so clean-burning that you could drink the effluent from the tailpipe with no urban smog from vehicles or generating stations.

For transportation, fuel cells have important advantages. Three main automotive goals are efficiency, range, and emissions. Gasoline and diesel fuels have the efficiency and range, but there are emissions problems. Batteries meet the emissions and the efficiency goals, but not the range. The fuel cell promises to have extremely low emissions, with excellent range and efficiency, providing the storage problems are solved. Hydrogen is an amazing substance. It is lighter than air. In its liquid form, you could throw it at people and it would evaporate before it hit them.

Fuel cells may be slow in coming. Fuel cell stacks are feasible in commercial form and the projected date of 2010 is a time DaimlerChrysler is comfortable with, after years of research and development. Complex fuel processors that can handle gasoline, like the system developed by Arthur D. Little, have been proved to work, but actual production models are still being developed.

DaimlerChrysler efforts to make a gasoline reformer work have been somewhat disappointing. The carmaker announced it would concentrate its efforts on methanol, signing on to the program advanced by its partners in the Alliance. DaimlerChrysler has also showed off a gasoline-powered fuel cell Jeep.

DaimlerChrysler has been slow on hybrids, since it is thought that Americans would not be excited about fuel savings. The Toyota Prius and other hybrids have proved this wrong and manufacturers could not keep up with the volume of demand. Until the merger with Daimler-Benz, Chrysler was working on both fuel cell and hybrid technology.

The design innovations in the ESX2 should be useful in future production versions. Getting a gasoline reformer to work efficiently in a production car will not be an easy task, although that would provide a path for practical fuel cell cars. DaimlerChrysler will sell the EPIC minivan and it has built several of car and bus fuel cell prototypes, starting with a hydrogen-powered internal-combustion minibus in 1975.

In the 1980s, before its work on fuel cells Daimler-Benz was experimenting with hydrogen in internal-combustion engines. It conducted road tests in Berlin from 1984 to 1988, with ten vehicles and over 350,000 miles of driving tests. In this same time period, BMW began testing cars that use liquid hydrogen and this work continues to this day. The German government committed more than $100 million to these projects.

But, burning hydrogen is less desirable than using it in a fuel cell.

The combustion of hydrogen releases carbon monoxide, hydrocarbons and some particulates, although these are about 0.1 of that from the burning of fossil fuels.

NECAR

Daimler-Benz built the NECAR (New Car) I, a commercial van that was its first fuel cell vehicle, in 1994. NECAR I was a prototype and most of the cargo area was used for the fuel cell equipment. The roof held a large hydrogen tank.

NECAR II was a smaller van built in 1996. It has seating for six and was capable of 60 miles per hour and could travel 150 miles before the onboard hydrogen tanks needed to be refilled. The range of 150 miles pushed Daimler-Benz into fuel reforming as seen on NECAR III. This has an onboard reformer and the range increased to over 300 miles.

NEBUS appeared in 1997 and showed the downsizing done by Ballard. It has ten of the company's 25-kilowatt fuel cell stacks in its rear compartment. It is a functional city bus, with a comparable range. It is similar but not identical to the buses Ballard has put on the streets of Vancouver and Chicago.

In 1998, Daimler-Benz made an important breakthrough and unveiled the world's first methanol-reformed fuel cell car. Daimler used its subcompact A-Class as a test bed. NECAR III was heavy and the fuel cell stacks and reformer took up everything but the front seats. Unlike the hydrogen-fueled cars, it did not perform smoothly, but it proved the concept that a methanol fuel cell car was feasible.

A joint venture called Daimler-Benz-Ballard (DBB) should help DaimlerChrysler to have production-ready cars, with 40,000 fuel cell cars possible in the first year.

Initially, pure hydrogen gas could be used for fleet vehicles, which includes delivery trucks, taxis and buses and onboard reformed methanol would be the fuel for passenger cars. Fleet vehicles are usually served by large garages with trained staff and could have the facilities for in-house hydrogen production. Without the reformers, fleet vehicles could be less complex and would be able to work within the 250-mile range limitation of onboard hydrogen tanks.

NECAR III was their first methanol vehicle and the reforming uses a four-step process, so there is some hesitation when you accelerate and some noise from the compressor. The hesitation problem may be solved with an injection system that mixes methanol and water. The fuel cells are two Ballard 25-kilowatt stacks since it is a small car.

Fuel cell research has become a major international trend with many engineers working on this technology worldwide. Germany already has enough methanol production to fuel 100,000 cars and worldwide, there is enough methanol for 2 million cars.

DaimlerChrysler has renewed its interest in liquid hydrogen, and that was the fuel in NECAR IV that appeared in 1999. NECAR IV was among the first drivable, zero-emission, fuel cell cars in the United States along with the Ford and GM fuel cell prototypes. It was a major advance over NECAR III, whose cell and reformer took up all the passenger space. NECAR IV is still heavy and slower to accelerate than Ford's P2000, but it has room for five, with a 40% power increase over the earlier version, a higher top speed of 90 miles per hour and a range of 280 miles.

Liquid Hydrogen

BMW will offer a 7 series version which will operate on hydrogen and gasoline. DaimlerChrysler and BMW are in partnerships with the German company Linde, which builds liquid hydrogen refueling stations. But handling liquid hydrogen is difficult, since hydrogen reaches a liquid state at minus 400°F, the cold fuel can cause serious damage to skin.

Liquid hydrogen stations could be run by robots. There already is such a station in Munich. Making liquid hydrogen work requires some new techniques, like attaching the tank to the car with a magnetic holder to isolate it from thermal convection.

A liquid hydrogen tank could be a little larger than a gasoline tank and it would offer a comparable range. A superinsulated tank can keep liquid hydrogen cold for weeks, but after a time it would warm up and return to a gaseous state, requiring that it be vented from the tank. This hydrogen gas might be recaptured and reused but this is one the challenges being worked on.

FCX

Toyota and Honda have been experimenting with both methanol and metal-hydrid storage of hydrogen. Honda has been building several test cars, in 1999 a Honda FCX-V1 (metal-hydrid hydrogen) and FCX-V2 (methanol) were tested at a track in Japan. The Ballard powered version-1 was driver ready, but proved to be a little sluggish and noisy. The other car suffered from a noisy fuel cell. Both Honda fuel cell test cars were built from the chassis of the discontinued EV Plus battery electric, although

Honda used a different and more aerodynamic body.

The FCX-V2 uses a Honda designed fuel cell and reformer. Miniaturizing an efficient methanol reformer remained to be done since both test cars had room only for a driver and passenger. Fuel cell equipment took up the rear seats. The need to test fuel cell cars under real-life conditions is one reason Honda joined DaimlerChrysler in the California Fuel Cell Partnership. Honda recently announced the first lease of its advanced FCX fuel cell vehicle.

Toyota's work with PEM cells began in 1989. It produced a methanol reformed car, the FCEV, in 1997. It was based on its electric RAV4.

Toyota has also worked with storing hydrogen in metal-hydrids. This technology has been tried by other companies and rejected because the metals are too heavy. Toyota obtained a 155-mile range with metal-hydrid storage. Toyota has also developed 35-MPa and 70-MPa high pressure hydrogen tanks that have been certified by the High Pressure Safety Institute of Japan.

Toyota will share some of its fuel cell research information and data from some of its environmental initiatives, including vehicle recycling and reduction of greenhouse gases with GM, its partner on a number of projects. GM will be spending $44 million in a joint project with the Department of Energy to put fuel cell demonstration fleets on the road in Washington, D.C., New York, California and Michigan.

In July 1998, Toyota said it would try to have a fuel cell automobile ready by 2003, but later this target date was dropped. Toyota was sharing technology with partner GM in a 5-year collaboration on electric, hybrid, and fuel cell cars. In 1998, the research division was testing methanol reformers and metal-hydrid hydrogen storage and had prototypes of each.

Toyota believed that there are major cost problems for onboard reformers and saw direct hydrogen as a big technical challenge. Still, it kept working in these areas and its FCHV (fuel cell hybrid vehicle) became the first vehicle in Japan to be certified under the Road Vehicle Act.

Volvo has modified a Renault Laguna station wagon with a 30-kilowatt fuel cell, running on liquid hydrogen. The Fuel Cell Electric Vehicle for Efficiency and Range (FEVER) car was partly financed by the European Union. It was completed in 1997 and has a 250-mile range. Even though it was a station wagon, the fuel cell car has room only for its driver.

Volvo is also a partner in another European project. This is the Capril project, which is managed by Volkswagon. A fuel cell VW Golf was built to run on methanol. Volvo developed the compressor, power converter and energy control system.

In 1992, Volvo built an aluminum-bodied hybrid Environmental Concept Car (ECC) to California emission mandates. It had the recyclable plastic panels and water-based paints that are used in a Volvo.

A series hybrid drive train was used where a diesel gas turbine drives a generator to charge a battery pack which powers an electric motor. The system is complex, but the car achieves good performance with low emissions and a 400-mile range. Volvo was testing 50-kilowatt stacks from Ballard before the Ford purchase.

Volvo has been building bifuel natural gas and gasoline cars and hybrids. By 1998 it was selling 500 bifuel sedans a year with many going to natural gas utilities in Europe. By 1999, it was working on a powersplit hybrid car, which automatically shifts from the electric motor to an internal combustion engine.

Hydrogen Infrastructure

A hydrogen infrastructure could cost hundreds of billions, since there is such a limited hydrogen-generating and distribution system now. Decentralizing production, by having reformers in commercial buildings and even in home garages in combination with local power generation would reduce some of the cost. Larger reformers in neighborhood facilities could be the service stations of tomorrow.

By 1998 the auto industry moved from weak commitments to a solid move toward fuel cells and EVs. All the auto companies were pursuing hydrogen fuel cells in some way. DaimlerChrysler is delivering fuel cell vehicles to customers in California. At the 2005 International Conference and Trade Fair on Hydrogen and Fuel Cell Technologies there were more than 600 fuel cell vehicles. In Europe the potential market for hydrogen and fuel cell systems is projected to reach several trillion Euros by 2020.

The new cars on the road in the future are likely to be a mix of vehicles including those with electric drive. This would include battery EVs, hybrids with direct-injection diesels, turbo generators and fuel cells.

Fuel Cell Cabs

London's first fleet of fuel cell taxis went into operation in 1998. The ZEVCO Millennium vehicle looks like a standard London taxi, but it has an alkaline fuel cell while most carmakers use PEM technology. The fuel cell charges a battery array used to power the electric motor. The fuel cell runs on hydrogen gas stored under the cab's floor and acts more like a range extender than a primary power source.

Shell Oil has established a Hydrogen Economy team dedicated to investigate opportunities in hydrogen manufacturing and fuel cell technology in collaboration with others, including DaimlerChrysler. One factor in the shift to fuel cells is concern over climate changes. Global warming is a factor of concern.

The world population is continuing to grow rapidly and the developing economies are starting to demand private cars. This creates more fuel demands and more urgency on environmental fronts and alternative fuels. International and American industry has been working on fuel cells and alternative energy.

For fuel cells to become commercially viable, the manufacturers must be convinced that they can make money on them. Los Alamos worked with General Motors and Ballard on PEM research which resulted in a 10-kilowatt demonstration unit.

DOE provides support to American companies, but the level of support has been less than the federal support in Germany and Japan. In 1993, Japan started a major 28 year, $11 billion hydrogen research program called New Sunshine. It surpassed Germany's hydrogen program to become the biggest program at that time. The basic hydrogen research included work on the metal-hydrid storage systems that are used in Toyota's fuel cells. German government support has declined since reunification. About $12 million was budgeted in 1995.

Direct Hydrogen Storage

Direct hydrogen research has included tests with fuel tanks pressurized at 5,000 pounds per square inch, which could provide a reasonable range without a reformer. Carbon-fiber composites which have been used in lightweight car bodies could also be used but they are very expensive.

The tank would need to provide a range of about 350 miles without using too much space. A light fuel cell car with a 5,000-psi carbon-fiber tank might be able to travel almost 225 miles before needing to be refueled.

Direct hydrogen storage is closer to an acceptable range, but there could be a liquid fuel stage. This would allow the use of the existing gasoline refueling infrastructure that cost hundreds of billions to build.

In 1998 a report prepared for the California Air Resources Board (CARB) called Status and Prospects of Fuel Cells as Automotive Engines appeared. The report favored methanol fuel cell stacks in cars over a direct-hydrogen infrastructure. Hydrogen is not as ready for private automobiles because of the difficulties and costs of storing hydrogen on board and the large investments that would be required to make hydrogen more available.

The report concluded that the automotive fuel cell is coming due to the almost $2 billion international investment. Fuel cells would provide an environmentally superior and more efficient automobile engine. This is being pursued with a combination of resources by strong organizations acting in their own interests and with support from public policy groups.

The report noted that hydrogen, even compressed at 5,000 pounds per square inch, may not be able to supply the required range. In one study by Ford, even with a fuel efficiency of 70 miles per gallon, the size of the tank needed for a 350-mile range would greatly impact both the passenger and cargo space. One alternative is to place the tank on the roof, like the NECAR II van, but this is not acceptable for a passenger car. Storing the fuel in special structures has been demonstrated by Toyota and Honda, but the metals are costly.

Northeastern University has worked on a system based on the absorption powers of carbon nanofilters for the high-density storage of hydrogen. This form of storage could make direct-hydrogen cars practical. The National University in Singapore has had some promising results in this area.

The Argonne National Laboratory reported that building a production capacity in the United States could cost $10 billion by 2015 and $230 to $400 billion by 2030. Building the distribution system could add $175 billion by 2030.

Directed Technologies is a consultant to Ford. They have stated that hydrogen could be delivered at around the same cost as its equivalent in gasoline, but these figures compare a 24.5-miles-per-gallon gasoline car using taxed gas with an 80-miles-per-gallon fuel cell car using untaxed hydrogen. If both vehicles get 80 miles to the gallon and neither fuel is taxed, hydrogen could cost 2 to 3 times more per mile. Generating hydrogen through renewable sources could reduce these costs.

The CARB report indicated that hydrogen would be produced at large, central facilities similar to a gasoline refinery. But hydrogen could be made at neighborhood refueling stations or at renewable energy farms.

One Princeton study of the Los Angeles area indicated the potential for solar photovoltaic plants in the desert areas east of the city. Enough hydrogen could be produced with solar power in an area of 21 square miles to fuel one million fuel cell cars. The wind site areas at Tehachapi Pass and

San Gorgonio are believed to have a similar potential. Geothermal power would be another renewable source. A problem in generating hydrogen this way is the long-distance pipelines required since the gas is leaky compared to other products.

The first commercial fuel cell cars may run on liquid fuels. The time tables announced by government and industry have generally been proven too conservative. Many auto companies already have running drivable fuel cell prototypes and it appears likely that some modest commercialization will be achieved in the next decade. Hybrids are already here and proving to be popular because of their efficiency.

Methanol Fuel Cells

Daimler-Benz has accumulated data on NECAR III emissions with a dynamometer programmed for a mix of urban and suburban driving. The results were promising since there were zero emissions for nitrogen oxide and carbon monoxide, and extremely low hydrocarbon emissions of only .0005 per gram per mile. NECAR III did produce significant quantities of carbon dioxide similar to the emissions of a direct-injection diesel engine where the fuel is injected directly into the combustion chamber. Direct-injection produces less combustion residue and unburned fuel.

Building a methanol infrastructure may not be as difficult. Methanol can be produced from natural gas and can also be distilled from coal or even biomass. In the 1980s, methanol was popular for a brief time as an internal-combustion fuel and President Bush even discussed this in a 1989 speech.

However, methanol is highly toxic and while it has some emissions benefits it adds tangible amounts of formaldehyde to the air. The world methanol infrastructure is the equivalent of about 5% of U.S. gasoline consumption, but new sources could be built up quickly. A major manufacturer of methanol, Methanex has stated that it could build a $350 million plant in 3 years that could fuel 500,000 cars.

Methanol can be pumped in existing gas stations, but since the fluid is corrosive the pumps, lines, and tanks would have to be made of stainless steel. If there is a demand, the costs would likely be handled by private investors.

Sulfur

Another problem with reforming is the presence of sulfur in the catalysts used in PEM fuel cells. One technique is to use a zinc-oxide bed to trap the sulfur, but this adds to the cost, weight, and size of the reformer. Refineries could also produce a new grade of gasoline with low sulfur content.

Along with being a smog enhancer, sulfur affects the performance of internal-combustion catalytic converters in the same way as it affects fuel cells. Sulfur can increase emissions by 20%. California uses low-sulfur fuel and the concentration is about 40 parts per million. In the rest of the country it is about 350 parts per million. A national low sulfur standard is estimated to add five cents per gallon to gasoline.

If fuel cell cars run on gasoline, there is minimum disruption, but many predictions indicate that methanol will serve as a bridge to direct hydrogen. Early fuel cell cars may run on methanol, but rapid advances in direct-hydrogen storage and production may push any liquid fuel out.

References

Ashley, Steven, "On the Road to Fuel Cell Cars," *Scientific American*, Volume 292 Number 3, March, 2005.

Motavalli, Jim, *Forward Drive*, Sierra Club Books: San Francisco, CA, 2000.

CHAPTER 5

THE NEW TRANSPORTATION

Faced with ever increasing regulations on exhaust emissions, fore-casts of oil shortages and potential global warming by greenhouse gases, the motor vehicle industry and national governments have spent tens of billions of dollars over the last decade on a cleaner, more efficient technol-ogy to replace the century old internal combustion (IC) engine.

Fuel cell vehicles are seen as the best long term option. The hydrogen fuel cell vehicle beats alternatives, such as hybrid vehicles. These combine IC engines with electrochemical batteries and still require petrochemical fuels that exhaust carbon dioxide and pollutants.

FUEL CELL TRENDS

Fuel cells are due to become a big energy player. In the future, we will start to see that more of the cars on the road will be fuel cell vehicles. There are obstacles and challenges to that happening, but there does not seem anything insurmountable. The cost of materials is being reduced and high-volume manufacturing can bring production costs down.

The present hydrogen powered cars continue to strive to hold enough fuel to get the 300 mile driving range of today's IC cars. Hydrogen service stations are few, so refueling becomes a problem. Some 12,000 fuel stations in the hundred largest cities in the U.S. would put 70% of the population within 2 miles of fuel. At a cost of one million dollars per station, $12 bil-lion would be needed to provide a fuel infrastructure. This is about half of what it would cost to build the Alaska pipeline in 2005 dollars. Shell Hy-drogen is planning for the first fuel cell cars in 2010 with a surge between 2015 and 2025.

Technical and market challenges could delay the commercial success of the fuel cell car. Car manufacturers must raise onboard hydrogen stor-age capacity, cut the price of fuel cell drive trains and increase the power plants' operating lifetimes. Hydrogen fueling will be required in enough stations to allow drivers to enjoy a range comparable to diesel fuel.

Over 50 million tons of hydrogen is produced worldwide per year and this is enough to fuel 200 million vehicles. Hydrogen produced from natural gas in a two step reforming process presently costs $4 to $5 per kilogram which is the chemical equivalent of a gallon of gasoline. If hydrogen is produced from water, it would take 50 kW of power costing about $2.50 per kg of hydrogen at present utility rates. This does not include other costs such the physical plants, storage facilities and transportation.

Fueling Stations

Plug Power is a Latham, NY based manufacturer of stationary hydrogen fuel cell generator units for backup power. A hydrogen fueling station has been developed with the help of Honda. This station uses a small steam reformer that extracts hydrogen fuel from natural gas using steam. The steam reformer was half the size of the previous version.

Along with refueling vehicles, the system provides hydrogen into a fuel cell stack to produce electricity for buildings on the site, which are also warmed by the waste heat generated by the power unit.

The fuel dispensing pump is about the size of a washing machine. First, the car is grounded by attaching a wire to the vehicle. The fuel hose nozzle is inserted into the refueling port and locked in place. Filling the car's tank takes about five or six minutes. The unit produces enough hydrogen to refill a single fuel cell vehicle a day. In Torrance, California Honda has built a service station that splits water into hydrogen and oxygen using solar power.

There are many problems facing the development of a hydrogen infrastructure. There is no demand for cars and trucks with limited fueling options and little incentive to invest in a fueling infrastructure unless there are vehicles on the road. The global cost of a complete hydrogen transition over the next 30 years could cost from $1 to $5 trillion.

One study by GM estimated that $10 billion to $15 billion would be needed to build 11,700 new fueling stations. This would allow a driver to be within two miles of a hydrogen station in most urban areas and there would be a station every 25 miles along major highways. The urban hydrogen stations could support about one million fuel cell vehicles. Twelve billion dollars may seem like a lot, but in today's world, some cable companies are paying $85 billion for cable system installations.

Hydrogen filling stations like this are now scattered in Europe, the U.S. and Japan. These are the first prototypes of an infrastructure with about 70 hydrogen refueling stations operating worldwide. The California

Hydrogen Highway program has a goal of 200 stations along major high-ways in the state.

The National Academy of Sciences committee feels that the transition to a hydrogen economy may take decades. Challenges include how to produce, store and distribute hydrogen in adequate quantities at reasonable costs without producing greenhouse gases that may affect the atmosphere.

The extraction of hydrogen from methane generates carbon dioxide. If electrolysis is used for splitting water into hydrogen and oxygen, the electricity may be produced by burning fossil fuels which generates carbon dioxide. Hydrogen is also a leak-prone gas that could escape into the atmosphere and set off chemical reactions.

Using fossil fuels to make hydrogen can take more energy than that contained in the hydrogen. Research at the Idaho National Engineering and Environmental Laboratory and Cerametec in Salt Lake City have found a way to electrolyze water and produce hydrogen with less energy.

The higher production rate of hydrogen is possible with high-temperature electrolysis. An electric current is sent through water that is heated to about 1,000°C. As the water molecules break up, ceramics are used to separate the oxygen from the hydrogen. The hydrogen that is produced has about half the energy compared to the energy required for the process.

The U.S. currently produces 50 to 60 million tons of hydrogen a year. But, this hydrogen may not be pure enough for fuel cells. Many of the problems in fuel cell development have occurred from impurities in the industrial hydrogen purchased for fuel.

The building of a hydrogen infrastructure in the 21st century can be compared to the investment in railroads in the 19th century or to the creation of the interstate highway system in the 20th century.

Transportation in the hydrogen economy is the fuel cell vehicle. In fuel cell development, the high cost of precious metals has led to ways to lower the platinum content. Methods include raising the activity of the catalyst, so less is needed and finding more stable catalyst structures that do not degrade over time while avoiding reactions that can contaminate the membrane.

Researchers at 3M have been able to increase catalytic activity with nanotextured membrane surfaces that employ tiny columns to increase the catalyst area. Other materials include nonprecious metal catalysts such as cobalt and chromium along with particles embedded in porous composite structures.

Hydrogen Storage

A fuel cell vehicle must have enough hydrogen to provide a reasonable driving range. For a range of 400 miles, 5-7 kilograms of hydrogen may be required, current fuel cell prototypes hold about half this amount.

Typically, hydrogen is stored in high pressure tanks as a compressed gas at ambient temperature. There is much work on doubling the pressure capacity of 5,000 psi (pounds per square inch) composite pressure tanks. But, twice the pressure does not double the storage. Liquid hydrogen systems that store the fuel at temperatures below -253°C have been successful. But, almost one third of the energy available from the fuel is needed to maintain the temperature and keep the hydrogen in a liquid state. Even with thick insulation, evaporation and losses through seals results in a loss every day of about 5% of the total stored hydrogen.

Alternative storage technologies include metal hybrid systems where metals and alloys are used to hold hydrogen on their surfaces until heat releases it for use. They act like a sponge for hydrogen. ECD Ovonic, a part of Texaco Ovonic Hydrogen Systems has been active in this area. The hydrogen gas in the high pressure storage tank chemically bonds to the crystal lattice of the metal or alloy in a reaction that absorbs heat. The resulting compound is a metal hybrid.

Waste heat from the fuel cell is used to reverse the reaction and release the fuel. In 2005 GM and Sandia National Laboratories launched a four-year, $10-million program to develop metal hybrid storage systems based on sodium aluminum hybrid.

Metal hybrid storage systems can be heavy and weigh about 300 kilograms. Researchers at the Delft University of Technology in the Netherlands have found a way to store hydrogen in water ice, a hydrogen hydrate, where the hydrogen is trapped in molecule sized cavities in ice. This approach is much lighter than metal alloys.

In the past, hydrogen hydrates have been difficult to produce, since they require low temperatures and pressures in the range of 36,000 psi. Working with the Colorado School of Mines, the Delft group used a promoter chemical (tetrahydrofuran) to stabilize the hydrates at only 1,450 psi. This approach would allow about 120 liters (120 kilograms) of water to store about six kilograms of hydrogen.

GM is considering a system that would cool hydrogen to -196°C at 1,000 psi. This would be less costly and reduce the boil off. Storing in hybrids creates heat and in order to get the hydrogen out the hybrid must be heated.

Fuel Cell Freeze Up

In Albany, NY, the state government started leasing a few Honda FCX hydrogen fuel cell cars on a cold November morning. Previous fuel cell vehicle demonstration programs have occurred in warmer areas to ensure that the fuel cell stacks would not freeze up. Subzero temperatures can change any liquid water present into expanding ice crystals that may puncture thin membranes or crack water lines. Honda has demonstrated that their fuel cell units can operate under winter conditions, an important achievement for practical fuel cell cars.

The freeze-resistant 2005 FCX models can operate at -20°C. Other companies, including DaimlerChrysler and GM have also had success with cold-starting cells. The technique used is to keep all water present as a vapor and not allow water droplets to occur.

Fuel Cell Auto Technology

The 2005 Honda FCX version is a four-seat compact hatchback with an ultracapacitor to provide short bursts of power for passing and hills. Most of the other fuel cell vehicles use batteries for this power. The use of the ultracapacitor could eliminate the expensive replacement of the batteries when battery life is over. The energy from a regenerative braking system is stored in the ultracapacitor, which is a low voltage, high efficiency capacitor. The 2005 FCX has a top speed of 92 miles per hour with a range of 200 miles. Equivalent fuel economy is 62 miles per gallon for city driving and 51 on the highway.

The Peugeot Quark ATV uses an air-cooled fuel cell with a 40 cell nickel metal hybrid battery. The 9 liter hydrogen tank can be pressurized to 10,150 psi for a range of up to 80 miles. The tank is designed to be exchanged for a full one when empty. Each 17 inch wheel has its own electric motor that can produce 74 pound-feet of torque. The motors also supply regenerative braking.

The General Motors Sequel fuel cell concept car holds enough fuel for 300 miles. It fits the seven kilograms of hydrogen into an 11-inch thick skateboard chassis. The Sequel has been called a crossover SUV. Since mechanical components are replaced by electrical parts, interior layouts can be more open with more space in smaller vehicles. General Motors is also providing 13 fuel cell vehicles for evaluation in the New York City area.

DaimlerChrysler has a fleet of 60 fuel cell cars called the F-Cell for worldwide testing. They have also built 33 fuel cell buses for 10 European cities as well as Beijing and Perth. Ford is testing 30 of its Focus FCV fuel cell compacts.

A hydrogen-based economy could be the ideal scenario for personal transportation. Hydrogen has a lot of advantages but a major problem is the lack of the hydrogen infrastructure. Hydrogen can also be difficult to handle since it is flammable at 4% concentrations.

A decreased reliance on fossil fuels, with electrics, fuel cell cars, hybrids and much-improved internal combustion engines is possible. The internal combustion engine may not remain dominant, but it not clear what technology will replace it. As cost-competitive fuel cell stacks become available, it will change the competitive mix of vehicles and create more demand for hydrogen fuel. The present period is not unlike 1900, when gasoline, electric, and steam cars all competed for market share, with the public and the industry unsure of the future. The world population is continuing to grow rapidly and the developing economies are starting to move into private cars. This creates more fuel demands and more urgency on environmental fronts.

In the greater stream of historic events the age of fossil fuels could be a brief period, with all its problems, lasting a little more than a single century. Fifty years from now the fossil fuel era may be seen as a distant memory just like we view the steam age now.

Fuel Cell Advantages

For transportation, fuel cells have significant advantages. Three main automotive goals are efficiency, range, and emissions. Gasoline and diesel fuels have the efficiency and range, but there are emissions problems. Batteries reduce the emissions and have enough efficiency, but not the range needed. The fuel cell promises to have extremely low emissions, with excellent range and efficiency.

Hy-Wire

In GM's Hy-Wire hydrogen powered concept vehicle, there is a fuel cell for the power source and electronics replace mechanical parts in the steering and braking systems. The driver looks through a large, sloped windshield that covers space usually taken up by an engine. There is no dashboard, instrument panel, steering wheel or pedals, only a set of adjustable footrests.

All controls are electronic, the driver twists a pair of handles to go, moves them to turn and squeezes to stop. The car's fuel cell produces 94 kilowatts of power which is equivalent to 126 horsepower, about the same as a Ford Focus. The Hy-Wire generates a loud whine while moving and can travel 140 miles before refueling.

Individual drive motors on each of the vehicle's four wheels allows a fuel cell powered all wheel drive system. Three tanks hold Hy-Wire's hydrogen fuel, compressed at 5,000 pounds per square inch. These were developed by Quantum Fuel Systems, the company that developed the industry's first 10,000-psi tanks, which could allow a fuel cell car to have a driving range of 230 miles.

Beneath the passenger cabin is an 11-inch-thick aluminum frame that holds all of the electric motors, microprocessors, mechanical parts, fuel-cell components, hydrogen tanks and other systems needed to operate the vehicle. The control wiring is carried in a single harness and permits designers to locate the operating controls virtually anywhere in the wide-open interior.

The compact, flat profile of GM's fuel cell which is about the size of a personal computer frees designers from the structure imposed by making room for a large internal combustion engine.

In addition to GM, DaimlerChrysler, Ford Motor Company, Honda Motor Company, Toyota Motor Corp. and others have spent billions developing alternative-fuel vehicles. GM has vowed to become the first carmaker to sell a million fuel cell vehicles and expects to have them on the market by 2010.

A hydrogen infrastructure could cost hundreds of billions, since there is a limited hydrogen-generating capacity now. But, decentralizing production, by having reformers in buildings and even in home garages in combination with local power generation, reduces some of that excessive cost. Larger reformers in neighborhood facilities could be the gas stations of tomorrow.

Hydroelectric dams could also be impacted by fuel cells. With more fuel cells around, electricity prices may fall and dam owners could make more profit selling hydrogen than selling electricity.

One study of the near-term hydrogen capacity of the Los Angeles region concluded that hydrogen infrastructure development may not be as severe a technical and economic problem as often stated. The hydrogen fuel option is viable for fuel cell vehicles and the development of hydrogen refueling systems is taking place in parallel with various fuel cell vehicle demonstrations.

Hydrogen fuel cells are being prompted by the desire to reduce global warming and control the spread of pollution in the developing world.

Fuel cells offer a major step in improved efficiency and reduced emissions.

Modeling Fuel Cell Cars

Some computer models of fuel cell cars show how much power is needed at the wheel, computed from the weight of the car, energy of the fuel, accessories and other variables including mileage. An onboard reformer is shown to provide 70 miles per gallon, but compressed hydrogen increases this to the equivalent of 100 miles per gallon. The models show that 60-70 miles per gallon is possible in hybrid cars using small gasoline or direct-injection diesel engines, which have much higher emissions.

In 1998 the auto industry moved from weak commitments to a solid move toward fuel cells and EVs. All the auto companies are pursuing hydrogen fuel cells in some way. But, the new cars on the road in the near future are likely to be a mix of vehicles including those with electric drive, including battery EVs, hybrids with gasoline and direct-injection diesels, turbo generators and fuel cells.

The move to fuel cells may not be pushed by declining oil supplies. The cost of developing new oil discoveries continues to fall and we may not see a forced drop in productivity. It was thought that there was 1.5 billion barrels of oil in the North Sea, but now there appears to be 6 billion barrels. We may not begin to reach the physical limits of oil production until mid-century. But, supplies could tighten quickly from natural or man-made disasters.

One factor in the shift to fuel cells is concern over climate changes. Global warming is a factor that most people perceive. However, is global warming being affected by our carbon economy?

THE END OF THE CARBON ECONOMY

We may be looking at the end of the carbon economy and the replacement of internal combustion power with fuel cells. Technology is driving our lives with tiny chips that have many times the computing power of larger 3-year-old computers, yet they cost less to manufacture.

Lighter, stronger materials and structures make electrical drives more feasible. Technology, legislative mandates and increased competition for markets will drive the fuel cell for automobiles and electrical power.

PNGV

The Clinton administration started the PNGV (Partnership for a New Generation of Vehicles) program at the end of 1993. It had a long term goal of an environmentally friendly car with up to triple the fuel efficiency of current mid-size cars without sacrificing affordability, performance or safety. It was a national research program with research support for over 350 automotive suppliers, universities and small businesses.

Seven government agencies and the United States Council for Automotive Research (USCAR) joined with representatives from GM, Ford, and DaimlerChrysler. It was a ten year plan to produce low-emission, 80-per-gallon family cars. The first designs appeared as working models in 2000. There was a timetable calling for production-ready prototypes, but the program was not binding.

PNGV was designed to get the government and industry together for a clean-air program aimed at cutting dependence on foreign oil, on a budget of about $200 million a year. The program did not do much about fuel economy since the average car being scrapped today gets better gas mileage than the average car displayed in dealers' showrooms.

A 27.5 miles per gallon CAFE standard was set for the different product ranges in 1985 and never achieved. There have been only light penalties for producing the present mix with the large sport-utility vehicles, which are about 30% of new car purchases.

Import cars have also not increased fuel economy. In 1992, the American car buyer could purchase some 26 compact cars that got at least 30 miles per gallon, but this number has dropped to less than ten. In Europe, high gas prices and smaller roads force smaller cars to be purchased.

Proposals to increase CAFE standards have never gained congressional majorities. In 1999 31 U.S. senators sent a letter to President Clinton pushing for stronger standards.

Carmakers have not voluntarily increased the fuel efficiency of their automobiles. Internal combustion engines are designed when new to produce emissions that are lower than state and federal standards. Automakers compete for market share but their interests may diverge on design issues.

PNGV had the involvement of nineteen federal government labs. The Department of Energy (DOE) provided about two-thirds of the federal support for PNGV research and development efforts. In 2003, PNGV was transformed into the FreedomCAR program with the focus on research and development in fuel cells and hydrogen infrastructures and technologies.

The DOE now has a fuel cell automotive program and the government has a somewhat abstruse alternative fuels strategy. The DOE's role is to support the research that validates the technology. It is up to the automakers to use it in vehicles and put it in the dealer's showrooms.

The DOE has funded research in areas such as compressed natural gas storage, direct-injection diesels with emissions-reducing catalytic converters, direct-hydrogen fuel cells including a 50-kilowatt automotive unit than runs without an air compressor and the Epyx gasoline reformer. DOE also supports national laboratories, such as Los Alamos which has been working on PEM fuel cells. This is the type used on the Gemini space program.

The liquid fuel reformer that has been worked on at Los Alamos and the Argonne National Labs is a fuel-flexible processor which can reform gasoline, natural gas, methanol, or ethanol at the control of a switch. This would also allow the use of the existing fuel infrastructure, but this approach forces the use of a more complicated, heavier system.

Los Alamos worked with General Motors and Ballard on PEM research which resulted in a 10-kilowatt demonstration unit. For fuel cells to become commercially viable the manufacturers must be convinced that they can make money on them. DOE provides support to American companies, but the level of support is less than the federal support in Germany and Japan.

Government Support

In 1993, Japan started a major 28 year, $11 billion hydrogen research program called New Sunshine. It surpassed Germany's to become the biggest program at that time. The basic hydrogen research included work on the metal-hydrid storage systems that are used in Toyota's fuel cells. German government support has declined since reunification. About $12 million was budgeted in 1995.

Hydrogen fuel cells have become an international competition. The liquid fuel reformer that has been worked on at Los Alamos and the Argonne National Labs is a fuel-flexible processor which can reform gasoline, natural gas, methanol, or ethanol at the control of a switch. This would allow the use of the existing fuel infrastructure. But, the approach forces the use of a more complicated, heavier system.

Direct-hydrogen research has included tests with fuel tanks pressurized at 5,000 pounds per square inch, which could provide a reasonable range without a reformer. Carbon-fiber composites which have been in lightweight car bodies could also be used but they are very expensive. The tank should be able to provide a reasonable range without using too much space. A light fuel cell car with a 5,000-psi carbon-fiber tank could travel almost 250 miles before needing to be refueled.

Direct hydrogen storage is getting close to an acceptable range, but there may be a liquid fuel stage with reformers. This would allow the use of the existing gasoline refueling infrastructure that cost hundreds of billions to build.

The Argonne National Laboratory reported that building a production capacity in the United States could cost $10 billion by 2015 and $230 to $400 billion by 2030. Building the distribution system could add $175 billion by 2030.

In 1998 a report prepared for the California Air Resources Board called Status and Prospects of Fuel Cells as Automotive Engines appeared. The report favored methanol fuel cell stacks in cars over a direct-hydrogen infrastructure. Hydrogen is not as ready for private automobiles because of the difficulties and costs of storing hydrogen onboard and the large investments that would be required to make hydrogen more available. The report concluded that the automotive fuel cell is coming due to the almost $2 billion international investment.

Fuel cells would provide an environmentally superior and more efficient automobile engine. This is being pursued with a combination of resources by strong organizations acting in their own interests and with support from public policy groups.

The report noted that hydrogen, even compressed at 5,000 pounds per square inch, may not be sufficient to supply the required range. In one study by Ford, even with a fuel efficiency of 70 miles per gallon, the size of the tank needed for a 350-mile range would greatly impact both the passenger and cargo space. But, computer models show that 100 miles per gallon are possible.

An alternative is to place the tank on the roof, like the NECAR II van, this could be acceptable in a high roofed van but not in most passenger cars. Storing the fuel in special structures has been demonstrated by Toyota and Honda, but the metals and structures are costly.

Northeastern University has worked on a system based on the absorption powers of carbon nanofilters for the high-density storage of hydrogen. This form of storage could make direct-hydrogen cars practical. The National University in Singapore has had some promising results in this area.

Hydrogen Versus Gasoline

Directed Technologies is a consultant to Ford. They have stated that hydrogen could be delivered at around the same cost as its equivalent in gasoline, but these figures compare a 24.5-miles-per-gallon gasoline car using taxed gas with an 80-miles-per-gallon fuel cell car using untaxed hydrogen. If both vehicles get 80 miles to the gallon and neither fuel is taxed, hydrogen could cost 2 to 3 times more per mile.

Generating hydrogen through renewable sources could reduce these costs. The California Air Resources Board (CARB) report indicated that hydrogen would be produced at large, central facilities similar to a gasoline refinery. But hydrogen could be made at a neighborhood refueling station or at renewable energy farms. One Princeton study of the Los Angeles area indicated the potential for solar photovoltaic plants in the desert areas east of the city. Enough hydrogen could be produced with solar power in an area of 21 square miles (a 3 by 7 mile section) to fuel one million fuel cell cars.

The wind site areas at Tehachapi Pass and San Gorgonio are believed to have a similar potential. Geothermal power would be another renewable source. A problem in generating hydrogen this way is the long-distance pipelines required since the gas is leaky compared to other products.

The time tables for fuel cells announced by government and industry have generally been proven too conservative. Many auto companies already have running drivable fuel cell prototypes. There was also some modest commercialization being achieved by 2004. Hybrids are proving to be in demand and most manufacturers have models available or are planning to introduce hybrids into their lines.

If fuel cell cars run on gasoline, there is minimum disruption. Many have predicted that methanol will serve as a bridge to direct hydrogen. But, hydrogen fuel cell vehicles are appearing. Rapid advances in direct-hydrogen storage and production may push any liquid fuel out.

Shell Oil has established a Hydrogen Economy team dedicated to investigate opportunities in hydrogen manufacturing and fuel cell technology in collaboration with others, including DaimlerChrysler.

On-site Power

A hydrogen economy may be jump started with distributed power; stationary fuel cells that generate on-site power in critical areas, schools, apartment buildings or hospitals. The cost of fuel cells would have to come down to less than $1,000 per kilowatt. At $800 to $1000 per kilo-watt fuel cells would be economical for buildings. The excessive space and weight would not matter so much for an installation in a basement or in an outside area.

The waste heat the fuel cell generates can be used in a cogeneration process to provide services like heating, cooling, and dehumidification. Instead of the 50% efficiency of a fuel cell with a reformer, or 60-70% without one, you can reach 90% or better of the total system efficiency.

In most situations, the waste heat is enough of a commodity to pay for a natural gas line and a mass-produced reformer to turn it into hydrogen. Then, the effective net cost of providing electricity to the building is about 1-2 cents per kilowatt hour.

Cars as Power Plants

As the building market for fuel cells grows, costs will come down and allow more economical fuel cells in cars. Buildings use 2/3 of all electricity in the United States, so there is the possibility of large fuel cell volumes. Both the building and vehicular fuel cell markets are potentially so large that when either of them starts moving it will push the other.

Stationary and mobile fuel cells could have a potential relationship that goes beyond cost and volume. If you have a fuel cell in a vehicle, then you also have a multi-kilowatt power generator on wheels, which is driven about 5% of the time and parked the other 95% of the time.

These fuel cell cars could be used to provide power and even water to buildings where people live or even work. Commuters could drive their cars to work and connect them to a hydrogen line. While they worked, their cars would be producing electricity, which they could then sell back to the grid. The car, instead of just occupying a parking space, would become a profit generator.

Thinking about cars as power plants is not something that we are conditioned to do, but it is an indication of how fuel cells could impact our lives.

Mobile Utilities

Fuel cell vehicles could provide extra value when they are in use, by acting as these mobile power sources. Most cars are used for 1 or 2 hours of the day. When they are not used, they are often parked where electricity is needed, offices, stores, homes or factories. If all cars were a fuel cell powered, the total power generation capacity would be several times greater than the current U.S. power generation.

Parked fuel cell cars could be plugged into the grid to generate power and these transportation fuel cells would operate as stationary fuel cell power sources. If only a small percentage of drivers used their vehicles as power plants to sell energy back to the grid, many of the power plants in the country could be closed.

However, if the major source of hydrogen is reformed natural gas, the cost of generating electricity with a low-temperature fuel cell would be about $0.20 per kilowatt-hour. This is more than double the average price for electricity. It would also produce 50% more carbon dioxide emissions than the most efficient natural gas plants which are combined cycle natural gas turbines.

Low-temperature fuel cells operating on natural gas are not as efficient at generating electricity. A stationary fuel cell system achieves high efficiency by cogeneration.

But, cogeneration would add to the complexity of the vehicle. Connecting a vehicle to the electric grid will also require some additional electronics. Extracting useful exhaust heat would involve new ductwork and possibly heat exchangers. This could be a problem for existing buildings, where parking may not be adjacent to heating units. Most homes could probably use the heat from a 1 kilowatt (kWh) fuel cell, but a car will probably have a 60/80-kW fuel cell.

Home electricity generation with either a stationary or a mobile fuel cell may not provide any cost savings that would jump-start commercialization. Also, a method is needed to get hydrogen to your home or office to power the fuel cell after your car's onboard hydrogen is consumed. For relatively small amounts of hydrogen, bottled hydrogen is likely to be expensive per kilogram. It could also be expensive to generate hydrogen on-site. Hydrogen generation and purification units may be too expensive for home use and local electricity and natural gas prices are much higher than for larger users.

Transportation fuel cells are being designed for about 4,000 hours of use, which gives a car a 10-year lifetime since they are used only a small percentage of the time. But 4,000 hours represents less than half a year under continuous use for generating electricity. Fuel cells for power plants are designed for 40,000 hours or more.

To reduce the cost of proton exchange membrane (PEM) fuel cells for transportation to under $100/kW the membranes are very thin and are proving less durable than the bulky stacks used for stationary applications. Some prototype tests have indicated a lifetime of 1,000 hours.

Spinning Reserves

While transportation fuel cells may not be generating grid power on a continuous basis, intermittent generation during peak power periods might be utilized. During the hottest days of summer in many parts of the country, air conditioning demand is high and, power generation costs go up. This peak power generation represents a source of revenue, where peak prices are high.

One California study suggests that vehicle owners could contribute to spinning reserves. These are contracts for generating capacity that is synchronized with the power line and when required must come up to its full output within 10 minutes. Spinning reserves contribute to grid stability especially during peak periods.

Since cars are designed to start quickly, they could add their power to the electric grid when needed. Utilities would contract this service, provided that it could depend on the needed power. This would make it more practical for fleets or business users. Individual owners would not want to lose control over access to their cars and they would if they had to connect them to the grid whenever there is a utility demand. Given how rarely individual car owners are likely to be called on to make their cars available for power, it is unclear that installing grid interconnect in private homes would become popular because of the safety, cost and logistical issues that it would raise. Car manufacturers might specify in their warranty that fuel cell engines could be connected to the grid only tens of hours per year, to ensure that the vehicle's life-time was not compromised. Spinning reserves are more likely to be needed at late afternoon or early evening, when a commuter uses a car to get home.

A fleet user or a business could provide an incentive for its employees and retrofit its parking lot for grid connection. PEM fuel cell vehicles might be reduced in ownership cost in that value might be extracted by using them for spinning reserves. A million fuel cell vehicles would provide over 70,000 megawatts (MW), which is about 10% of total U.S. generation capacity. It is more than enough for spinning reserve. This could help subsidize fuel cell vehicles.

Besides using fuel cell cars as mobile utilities, stationary fuel cells can be used to cogenerate electricity, heat, and hydrogen. In high-temperature fuel cells, hydrogen can be separated and processed at low cost.

This can boost overall system efficiency to 90% in converting natural gas. Solid oxide fuel cells (SOFC) could provide hydrogen at a few dollars per kilogram.

Residential Power

Fuel cells could still become a provider of home power. Proton exchange membrane (PEM) fuel cells may become a part of an overall home hydrogen fueling system for fuel cell cars. The average American home requires about 1-kW of power per day with a peak of more than 4-kW. Most of the peak is due to large and small appliances, heating and air conditioning and power tools. A smaller amount is needed for lighting and electronic devices.

The load is higher in the summertime, when air conditioning is needed. Hot water demand is almost constant year-round. Heating demand peaks in the wintertime, along with lighting and electronics use.

A small fuel cell of about 1-kW in capacity, could cogenerate year-round to provide base-load power and hot water. Larger fuel cells would provide too much heat than a home can use most of the time. The DOE estimates that the optimal size for a residential combined heat and power (CHP) or cogeneration system in the United States is about 0.75-kW for a PEM fuel cell.

Small residential PEM fuel cells may be 35% efficient or less in converting natural gas to electricity. An averaging of all the CO_2 emissions from all the U.S. power plants, indicates that all the coal, nuclear, hydroelectric and other plants in the entire U.S. electric grid is roughly the equivalent of one mammoth 30% efficient natural gas power plant. So, unless the home PEM unit is cogenerating usable heat most of the time that it is operating, and thus replacing the natural gas (or other fuel) used to produce hot water, the system is not avoiding significant greenhouse gas emissions. The DOE's analysis found that the home PEM system would cut CO_2 emissions by less than 10%. A larger PEM would achieve even smaller savings, since a smaller fraction of the waste heat would be utilized. Most newer power plants use natural gas in a combined cycle for an efficiency over 55%.

Long before PEM fuel cells become inexpensive enough to compete with gasoline engine generators, which can cost $50 per kW, PEMs will become competitive with stationary power plants, which can cost $500/kW or more.

One big difference for stationary PEMs, however, particularly in homes, is that in the foreseeable future they will probably need a reformer to extract hydrogen from natural gas in order for hydrogen to be convenient and cost-effective as a consumer product.

It would not be very useful for homeowners to use their own electric power to electrolyze water for hydrogen to run a fuel cell to generate electricity. Delivery of hydrogen to the home is likely be prohibitive in the near future.

Developing a small, affordable reformer for delivering high-grade hydrogen for a PEM fuel cell has been difficult. Ballard, a fuel cell pioneer and a leader in PEM fuel cells, dropped its fuel reformer development. Ballard has a 1-kW PEM fuel cell for $6,000 that runs on pure hydrogen. The product is available with storage tanks for industrial and residential use.

Another fuel cell company, Plug Power, has a 5-kW PEM fuel cell system for backup power which sells for $15,000 ($3,000/kW). The system runs on hydrogen with a design life of 1,500 hours over 10 years. PEM fuel cell systems are available with a reformer for $10,000-$12,000/ kW.

Costs are likely to drop, but right now, many fuel cell and processor component costs are rigid and do not depend so much on the power output. In smaller systems, the cost per kilowatt is higher. While 20-kW units could cost $1,500/kW, a small 2-kW unit might cost $5,000/kW or more.

Installation costs are also likely to be high at this time. Initially, it will be a specialty product and sold like solar systems.

Fuel cell power systems could be integrated with the home's electric and heating systems in the future. Single-family PEM fuel cell units could eventually approach an installed cost of $1,000/kW in high-volume production.

The PEM system could be used with net metering. When the home system is generating more electricity than the home is using, power is returned to the electric grid and the electric meter runs backward. By 2003, almost half of the states had net metering for residential power systems.

Typically, consumers are only allowed to sell electricity back to the utility at retail rates only for electricity production that does not exceed their total annual consumption. For any excess production, the homeowner receives the rate that the utility pays to a large central power plant. This is the utility's avoided cost rate which is much lower than the residential rate. If your home consumed 10,000-kWh of electricity during the year, but your fuel cells generated 15,000-kWh, you would get paid at the retail rate for the first 10,000-kWh generated and a lower rate for the remaining 5,000-kWh.

Home fuel cell power may be utilized in those parts of the country where electricity prices are very high and in rural areas where there are electricity transmission and distribution problems.

In some urban areas, there are air quality and other restrictions that have made it difficult to build new generation facilities. Solid oxide fuel cells (SOFCs) may be a better technology for power generation with their higher electric efficiency and more usable heat. They do not need an external reformer.

Since they operate at very high temperatures, they take several hours to warm up and are more suitable in commercial and industrial applications that require high levels of electricity continuously.

Green Power

A recent trend is the willingness of some consumers and businesses to pay more for products that are perceived as green. Many companies in the United States are increasingly concerned with the environmental effects of their energy consumption. The World Wildlife Fund (WWF) and the Center for Energy and Climate Solutions, have been working with businesses such as IBM, Johnson & Johnson and Nike in the Climate Savers Program to develop and adopt innovative climate and energy solutions.

By 2010, the efforts of the first six Climate Savers companies may result in annual emissions reductions equivalent to more than 12 million metric tons of CO_2.

One basic technique for reducing emissions is to use energy more efficiently. This includes more efficient lighting or sources of power such as renewable solar sources and cogeneration. Stationary fuel cells could be a part of this mix.

Fuel Cell Incentives

The federal government offers a $1,000/kW subsidy to encourage the use of fuel cells. Several states, including New Jersey, Connecticut and California also offer subsidies. Some of these exceed $2,000/kW to reduce the cost of fuel cell systems. Other subsidies can also be important in the commercialization of stationary fuel cells.

Starwood Hotels and Resorts Worldwide installed two molten carbonate fuel cells in its Sheraton Parsippany and Sheraton Edison hotels in New Jersey with the aid of a $1.6 million grant by the New Jersey Clean Energy Program.

The two 250-kW fuel cells from FuelCell Energy each weigh almost forty tons and are about the size of a railroad car. The fuel cells are used to cogenerate electric power and hot water heating. They will provide about one fourth of the electricity and hot water needs for each hotel.

Starwood does not own or operate the fuel cells, PPL utility, in Pennsylvania has this responsibility, so there are no up-front costs, which makes this arrangement attractive to Starwood. It does not wish to be in the power generation business and it only put up about $40,000. PPL also guaranteed a lower electricity price than Starwood was currently paying as another incentive.

Starwood has a history of energy efficiency. It won the 2002 aware for Excellence in Energy Management for the hospitality sector from EPA's Energy Star program, which promotes practices in energy efficiency. In 2003, Starwood was the EPA's Energy Star Partner of the Year in the hospitality sector.

Another user of fuel cells is Dow Chemical. In 2003, Dow began installing 75-kW PEM fuel cells from General Motors in Freeport, Texas, Dow's largest chemical plant. Dow might use as much as 35 megawatts (MW) of power from the PEM generating system.

The fuel cells will run on hydrogen that is a by-product of the chemical manufacturing process. Excess hydrogen is also used in combustion engines at this facility to provide power. Although hydrogen can be more efficient than natural gas in combustion engines, the GM fuel cells may operate even more efficiently, with an electric efficiency of 45% or more. Dow will also use the cogenerated hot water from the fuel cells. The company cogenerates more than 90% of the power it uses.

Energy Efficiency

The total energy lost by U.S. power generators equals all the energy that Japan uses. The typical heating boiler converts only about 65% of the fossil fuel energy to useful heat or steam. Newer boilers can be more than 80% efficient. By generating electricity and capturing the waste heat in a cogeneration system, much energy and pollution can be saved. Overall fuel cell system efficiencies can exceed 80%.

Reliable Power

Highly reliable power is a major reason for on-site generation. The 200,000-square-foot Technology Center of the First National Bank of Omaha installed a fuel cell system in 1999. A power failure can cost one of its major retail clients $6,000,000 every hour in lost orders. The bank's system was made by SurePower of Danbury, CT. The bank installed the system to protect existing clients and to attract new business while gaining an edge over its competitors.

The typical uninterruptible power supply (UPS) generator/utility system has an availability of about 99.9%. This is the fraction of time that the system is available for use. A system available 99.9% of the time is called a 3-nines system, while a system with 99.99% availability is known as a 4-nines system. The best redundant UPS can achieve an availability of 4-nines. The fuel cell system at First National Bank of Omaha has an estimated availability of 6 to 7 nines.

A 4-nines system has a 63% probability of at least one major failure over a 20-year span. A 6-nines systems has only a 1% probability of at least one major failure in 20 years. When it was installed in 1999, the system was tested to have an availability in excess of 6-nines.

In the first four years, there was 12 power disturbances from the utility but the system operated flawlessly and the computers were not affected. The bank uses this experience as a main part of its marketing and has increased its market share.

The initial cost of the fuel cell system was higher than an UPS system, but the life-cycle costs show the fuel cell system to be less expensive.

The SurePower system uses four phosphoric acid fuel cells made by UTC Fuel Cells which provide 320-kW for critical power needs. The fuel cells are combined with flywheels and advanced electronics to minimize the affects of disruptions. Each of the four fuel cells provides 200-kW, so the system redundancy achieves the high availability.

Any excess electricity along with cogenerated heat is used by the bank and results in the lower life-cycle cost. Also, the fuel cell does not require an air-conditioned area like traditional UPS systems. This results in almost $30,000 saved in annual cooling costs. The system results in more than 40% lower emissions of CO_2 and less than 1/1,000 the emissions of other air pollutants.

To be cost-effective in a retrofit, new technology must compete against the sunk cost of all the old equipment, which is not an easy task for a new technology trying to break into the market at a relatively high price. But, in new construction, a fuel cell can compete with other power generating alternatives.

A new facility's energy system can be designed to utilize the waste heat effectively, capturing more of the efficiency a fuel cell can deliver. This can be more difficult in a retrofit. High-temperature fuel cells can run an absorption chiller, where the heat is used to provide cooled water for air-conditioning.

The need for high-reliability power has grown with the Internet and related digital hardware. In 1998 the Internet consumed 8% of U.S. electric power up from less than 1% in 1993. Computers and other office equipment consumed 13% of the country's electricity, up from 5% in 1993. The U.S. electric grid can be viewed as the equivalent of a huge 30% efficient natural gas power plant.

Fuel cells represent a new technology that must compete with alternatives such as more efficient boilers or cogeneration units or renewable energy. The efficiency of fuel cells can be as high as 80% or more when both the electrical power and heat energy are utilized. However, modern industrial boilers can also be 80% efficient or more.

A 1-year test in 2003 of a 3-kW stationary PEM fuel cell at a military facility indicated an overall electrical efficiency for running on hydrogen of 27%. As the technology improves, residential PEMs of several kilowatts running on natural gas are expected to be 35% efficient. Larger units for commercial and industrial use could exceed 40%. Cogeneration with a PEM fuel cell is not as effective since their waste heat may not be high enough for many industrial applications, which often require high-pressure steam. The exhaust temperature of PEM fuel cells is not high enough to run an absorption chiller, which uses heat in a refrigeration cycle. Phosphoric acid fuel cells operate at a slightly higher temperature than PEM cells for cogeneration.

Commercial office buildings in most regions of the country typically do not require heating for most of the year, and their hot-water needs are minor. But, many commercial office buildings require air conditioning in the summer and other times of the year due to internal heat generated by the occupants, equipment and lighting.

The SurePower system at the First National Bank of Omaha uses four PC24s and has more than 40% lower CO_2 emissions than the alternative of using the utility grid with a UPS.

A UPS is less efficient than the SurePower system since it requires more input power compared to delivered output power. It also requires a significant amount of power for cooling. The bank uses about 1/4 of the thermal energy provided by the fuel cells which results in an overall efficiency of almost 55%. High-temperature fuel cells can reach electric efficiencies of 50% and provide high-quality heat.

Molten carbonate fuel cells could reduce CO_2 emissions in industrial and commercial facilities by a third. Hybrid fuel cell and gas turbine systems could reduce these emissions by one half or more.

As stationary fuel cells reduce their costs with continuing R&D, they will be able to compete with other small- to medium-sized power generation sources for on-site generation, particularly cogeneration for factories and commercial buildings. The installed cost for fuel cell generation systems is expected to reach $800/kW.

Many studies indicate the potential is large. A 2000 study for the DOE's Energy Information Administration found that the total power needs for combined heat and power (CHP) at commercial and institutional facilities was 75,000-MW. Almost two thirds of these required systems of less than 1-MW.

These systems are a good match for fuel cell generation. The remaining power needs in the industrial sector are almost 90,000 MW. This does not include heat-driven chillers or systems below 100-kW.

Fuel cells have competition that is entrenched in very mature, reliable, low-cost technologies. Many barriers exist to impede the use of widespread use of small-scale CHP systems. These existing technologies and existing companies can be a formidable barricade against new technologies and new companies.

One reason renewable energy technologies do not have deeper market penetration in the United States today is not that those technologies fail to meet cost and performance goals. It is because the competition did not sit still and has been much tougher and inventive, than expected.

CHP Technology

A variety of technologies have been pursuing on-site CHP for years, including turbines, reciprocating engines, and steam turbines. Gas turbines in the 500-kW to 250-MW produce electricity and heat in a thermodynamic cycle known as the Brayton cycle. They produce about 40,000-MW of the total CHP in the United States. The electric efficiency for units of less than 10-MW, is above 30%, with overall efficiencies reaching 80% when the cogenerated heat is used.

They generate relatively small amounts of nitrogen oxides and other pollutants. Several companies have developed very low NO_x units. Their high temperature exhaust may be used to make process steam and operate steam-driven chillers. A 1-MW unit can cost $1,800/kW installed while 5-MW unit may cost $1,000/kW installed. In these systems, the turbine generator is about 1/3 of the total cost with the other costs including the heat recovery steam generator, electrical equipment, interconnection to the grid, labor, project management and financing.

Reciprocating engines are another mature product used for CHP. These stationary engines may be spark ignition gasoline engines or compression ignition diesel engines. Capacities range from a few kilowatts to over 5-MW.

Natural gas or other fuels may also be used in the spark ignition engines. Electrical efficiency may range from 30 for the smaller units to more than 40% for the larger ones. Reuse of the waste heat can provide overall efficiencies to 80%. The high-temperature exhaust of 700°F-1,000°F can be used for industrial processes or an absorption chiller. About 800-MW of stationary reciprocating engine generation is installed in the United States.

Development has been closely tied to automobiles and in the last few decades increases in electric efficiency and power density have been dramatic as well as emission reduction. Some units can even meet California air quality standards when running on natural gas.

Hess Microgen has a unit that can provide electricity, absorption cooling and hot water. The Starwood hotel chain is using Hess Microgen units in its hotels.

A 100-kW reciprocating engine generating system may cost $1,500/kW installed, while an 800-kW unit can cost $1,000/kW. The engine cost is about one fourth of the total price. The rest of the cost comes from the heat recovery system, interconnect/electrical system, labor, materials, project management, construction and engineering.

Steam turbines are an even older technology, providing power for over 100 years. Most utility power is produced by steam turbines. The steam turbine generator depends on a separate heat source for steam, often some type of boiler, which may run on a variety of fuels, such as coal, natural gas, petroleum, uranium, wood and waste products including wood chips or agricultural by-products.

Steam turbine generators range from 50-kW to hundreds of megawatts. In 2000, almost 20,000-MW of boiler and steam turbines were used to provide CHP in the United States. For distributed generation, a boiler and steam turbine system may be expensive.

But, a process that already uses a boiler to provide high pressure steam can install a back pressure steam turbine generator for low cost, high efficiency power generation. The pressure drops in the steam distribution systems are used to generate power. This takes advantage of the energy that is already in the steam.

A back-pressure turbine is able to convert natural gas or fuels into electric power with an efficiency of more than 80%, which makes it one of the most efficient distributed generation systems. The CO_2 emissions are low as well as pollution emissions. The installed capital cost for these systems is about \$500/kW. High efficiency, low cost and low maintenance allow these back-pressure installations to have payback times of two or three years.

Competition to a CHP project also includes price breaks from the local utility. When the local utility learns that a company is considering cogeneration, it sometimes offers a lower electricity rate in return for an agreement not to cogenerate for a certain period of time. This is especially true for bigger projects or those that might replace a large portion of its total load with on-site generation. A lower utility bill reduces the future energy cost savings from the CHP project and thus reduces the return on investment and increases the payback time.

Other barriers to distributed energy projects besides costs include project complexity and regulations. A 2000 report by the National Renewable Energy Laboratory, studied sixty-five distributed energy projects. The report found that numerous technical, business practice, and regulatory barriers can block distributed generation projects from being developed. These barriers include lengthy approval processes, project-specific equipment requirements and high standard fees.

There is no national agreement on technical standards for grid interconnection, insurance requirements or reasonable charges for the interconnection of distributed generation. Vendors of distributed generation equipment need to work to remove or reduce these barriers.

Starwood faced utility efforts in 2003 to block the installation of a 250-kW molten carbonate fuel cell in a New York hotel. It overcame these efforts mainly because the system represented only 10% of the hotel's total power. These barriers have been described as a battle between distributed generation and the local utility.

Distributed projects are not always given the proper credit for their contributions in meeting power demand, reducing transmission losses and improving environmental quality.

Fuel cells should have several advantages that will minimize some of these barriers, especially those related to environmental permitting. The federal government and some states, including California, are attempting to reduce these barriers and give fuel cells and other technologies credit for reducing strain on an overstressed electric grid. But, removing all these barriers can require years.

Japan has a target of 10 gigawatts of fuel cell generated power along with 5 million fuel cell vehicles by 2020. U.S. targets of 8 million fuel cell vehicles along with 20 gigawatts of fuel cell power are being promoted by this time period. This would be about 2% of U.S. installed capacity. A target of 20% of personal power markets is also being promoted. This would include battery powered items such as phones, computers and tools.

In the history of hydrogen, the contributions made by NASA (National Aeronautics and Space Administration) to develop compact, lightweight engines and energy storage devices for space applications have been important.

Hydrogen Generation

Hydrogen production has commercial roots that go back more than a hundred years. Hydrogen is produced to synthesize ammonia (NH_3), for fertilizer production, by combining hydrogen with nitrogen. Another major use is hydro-formulation, or high-pressure hydro-treating, of petroleum in refineries. This process converts heavy crude oils into engine fuel or reformulated gasoline.

Annual worldwide production is about 45 billion kilograms or 50 (kg) 0 billion Normal cubic meters (Nm^3). A Normal cubic meter is a cubic meter at one atmosphere of pressure and 0°C. About one half of this is produced from natural gas and almost 30% comes from oil. Coal accounts for about 15% and the rest 4-5% is produced by electrolysis.

Hydrogen production in the United States is currently about 8 billion kg (roughly 90 billion Nm^3). This is the energy equivalent of 8 billion gallons of gasoline. Hydrogen demand increased by more than 20% per year during the 1990s and has been growing at more than 10% per year since then. Most of this is due to seasonal gasoline formulation requirements.

In 2000, the U.S. consumed almost 180 billion gallons of gasoline, diesel fuel and other transportation fuels for road travel. This is about 20% of total U.S. energy consumption. When travel by air, water, and rail is added, including pipelines energy, total transportation energy rises to almost 30% of U.S. energy consumption.

References

Alley, Richard B., "Abrupt Climate Change," *Scientific American*, Volume 291 Number 5, November 2004, p. 64.

Kasting, James F., "When Methane Made Climate," *Scientific American*, Volume 291 Number 5, July 2004.

Schneider, Stephen Henry, *Global Warming*, Sierra Club Books,: San Francisco, CA, 1989.

CHAPTER 6

FUELS AND THE ENVIRONMENT

The spreading of economic development to all reaches of the globe has fueled the growth of the automobile. Most industrialized nations including Japan, Britain, Germany, France and others have seen great changes in energy growth as well. But by the end of the 20th century, the United States used more energy per capita than any other nation in the world, twice the rate of Sweden and almost three times that of Japan or Italy. By 1988, the United States, with 5% of the earth's population, consumed 25% of all the world's oil and released about a quarter of the world's atmospheric carbon.

CARBON ACCOUNTING TRENDS

Almost 4.5 billion years ago, the earth was formed, and 95% of the atmosphere consisted of carbon dioxide. The appearance of plant life changed the atmosphere since plants, through the process of photosynthesis, absorb carbon dioxide. Carbon from the atmosphere was drawn into the vegetation. When the vegetable matter died, it decomposed, and formed coal and oil. This reduced the carbon dioxide in the atmosphere to less than 1%.

Industrialization and the burning of fossil fuels reverses this process. Instead of being drawn out of the air, carbon is extracted from the ground and sent into the atmosphere. In the United States, a major surge in energy consumption occurred between the late 1930s and the 1970s, rising by 350%. More oil and natural gas was used to meet industrial, agricultural, transportation and housing needs.

Oil and natural gas contain less carbon than coal or wood, but the demand for electricity and fuel soared as the nation's economy grew and consumers became more affluent. By 1950, Americans drove three-quarters of all the world's automobiles and they lived increasingly in energy consuming suburban homes, with inefficient heating and cooling systems. Appliances were also inefficient. A 1970s era color television operated for four hours a day was the energy equivalent of a week's worth of work for a team of horses.

U.S. energy consumption slowed down in the 1970s and 1980s, as manufacturers introduced more efficient appliances. Even so, by the late 1980s, Americans consumed more petroleum than Germany, Japan, France, Italy, Canada and the United Kingdom combined.

THE INTERSTATES

As private cars began to control American transportation, they needed new roads to run on. In the early 1930s, the National Highway Users Conference, also known as the highway lobby, became one of the most influential groups in Washington. In years following World War II, interstate highways were to become the realization of the American dream.

GM President Charles Wilson, became Secretary of Defense in 1953, and proclaimed that a new road system was vital to U.S. security needs. Congress approved the $25 billion Interstate Highway Act of 1956 and the highways expanded.

The interstates encouraged more single-family homes which had to be reached by private cars. Since the mid-1950s, cities like Phoenix, Arizona, have grown from 15-20 to over 200-400 square miles. From 1970 to 1990, the greater Chicago area grew by more than 46% in land area, but its population increased by only 4%. The greater New York City area grew by 61% from 1965 to 1990, while adding only 5% to its population.

The American Automobile Manufacturers Association which merged into the International Alliance of Automobile Manufacturers claims that today's automobiles are up to 96% less polluting than cars 35 years ago but automobiles still produce a quarter of the carbon dioxide generated annually in the United States.

A global accord on reducing hydrocarbon emissions was reached at the 1992 Early Summit in Brazil. Great Britain and Germany came close to meeting their 2000 targets. The United States fell short of its goal by 15 to 20%. Could this effort at international cooperation succeed, when cars are the major culprits and almost every country is filling its roads with more and more of them?

The international agreement on global warming signed by 150

countries in Kyoto, Japan late in 1997 requires a drastic reduction in automobile exhaust emissions. The world's largest producer of carbon dioxide emissions is the United States. Greenhouse gases were to be reduced to 7% below 1990 levels by 2012.

There has been much opposition to Senate ratification of the Kyoto accords. Meeting the Kyoto goals would not be easy, the auto industry would have to do its part and many believe the economy would suffer greatly for reasons not understood completely. Developing countries like China and India would be exempt from the reduction of carbon dioxide emissions.

Tightening of the corporate average fuel economy (CAFE), the federal standard for cars and trucks, would be needed. A 12-mile-per-gallon car or truck emits four times as much carbon dioxide as a 50-mile-per-gallon subcompact. The auto industry has been against attempts to tighten CAFE. Vehicles like SUVs and pick-up trucks were not subject to CAFE standards.

The result was that average fuel economy has declined since 1988, as car manufacturers produced less fuel-efficient smaller vehicles and more profitable trucks and SUVs.

THE GREENHOUSE EFFECT

The greenhouse effect is a term used to describe the increased warming of the earth's surface and lower atmosphere due to increased levels of carbon dioxide and other atmospheric gases. Like the glass panels of a greenhouse, the effect lets heat in but prevents some of it from going back out. If it were not for the greenhouse effect, temperatures at the earth's surface today would be about 35°C (60°F) colder than they are, and our world would be much different.

There is some debate if the amount of the greenhouse gases will soon be increased by human actions to levels that are harmful to life. A rise in temperature of about 5°C (9°F) in the next five decades would be a rate of climate change about ten times faster than the observed average rate of natural change. These temperature changes could alter patterns of rainfall, drought, growing seasons and sea level.

The greenhouse effect is a widely known theory. It is based on the following arguments. The sun is the main source of the earth's climate with surface temperatures of about 6,000°C (10,800°F). This produces radiant energy at very short wavelengths. Almost half this energy reaches the earth's surface. Particles and gases in the earth's atmosphere absorb about 25% of this energy and 25% is reflected back to space by the atmosphere, mostly from clouds. About 5% of the incoming solar radiation is reflected back to space from the surface of the earth, mostly from bright regions such as deserts and ice fields.

A 1-square-meter surface (39 inches by 39 inches), placed above the atmosphere can collect about 1,370 watts of radiant power. This is called the solar constant even though it varies by a few tenths of a percent over 11-year-long sunspot cycles.

Since the sun is not shining all the time on every square meter of the earth, the total amount of incoming energy is about 340 watts per square meter.

Since the earth has temperature, it emits radiant energy known as thermal radiation or planetary infrared radiation. Satellites have measured an average radiant emission from the earth of about 240 watts per square meter. This is the radiation a black body gives off if its temperature is about −19°C (−3°F). This is also the same energy rate as the solar constant averaged over the earth's surface minus the 30% reflected radiation. Thus, the amount of radiation emitted by the earth is closely balanced by the amount of solar energy absorbed. Because the earth is in this state of radiation equilibrium, its temperature changes relatively slowly from year to year.

If more solar energy is absorbed than infrared radiation emitted, the earth would warm up and a new equilibrium would form. But, if somehow the earth were brighter (more clouds), it would reflect more solar radiation and absorb less. This would cool the planet and lower the amount of infrared radiation escaping to space to balance the lower amount of absorbed solar energy.

The earth's radiant energy balance today is 240 watts per square meter. The amount of energy the earth absorbs from the sun is the same amount it radiates back to space on average over the 500 trillion square meters of surface area. When satellites were able to get above the earth's atmosphere and measure the outgoing thermal radiation, it showed this balance to a high degree of precision.

An average of the temperature records on the earth's surface over a year shows that the earth's average surface temperature is about 14°C (57°F). But, the earth's 240 watts per square meter of thermal infrared radiation as measured by satellite are equivalent to the radiation emitted by a black body whose temperature is about -19°C (-3°F), not the 14°C (57°F) average measured at the earth's surface. This 33°C (60°F) difference between the apparent temperature of the earth as seen in space and the actual temperature of the earth's surface is attributed to the greenhouse effect.

The solar heat absorbed by the atmosphere and the earth's surface goes back into the atmosphere through the evaporation of water, thermals formed by the heating of air in contact with a warm surface and the upward emission of energy.

The trace gases in the earth's atmosphere are only a few percent of its composition but they make the planet habitable. They absorb radiant energy at infrared wavelengths much more efficiently than they absorb radiant energy at solar wavelengths, thus trapping most of the radiant heat emitted from the earth's surface before it escapes.

Greenhouse Gases

These gases are mostly water vapor, carbon dioxide (CO_2) and particles, mostly water droplets in clouds. They absorb infrared energy and also give it off. This infrared radiation is emitted upward cooling the planet and maintaining a balance with incoming sunlight. Some goes back to the earth's surface creating the greenhouse effect. The downward reradiation warms the earth's surface and makes it 33°C (60°F) warmer.

The greenhouse analogy says that the gases and clouds in the earth's atmosphere let a larger amount of the sun's shorter wavelength radiation in and allow the longer wavelength infrared radiation to escape to space. This theory has come into being due to millions of measurements in the atmosphere, space and laboratories.

About 4.5 billion years ago the heat from the sun was about 30% less powerful than it is today. A number of elements, including carbon and oxygen, condensed to form the earth. As the earth's crust cooled and hardened, hot gases from the interior were ejected including carbon dioxide.

The amount of carbon dioxide in the atmosphere then has been estimated to be many times greater than today. This explains how the earth's climate was warm enough for liquid water and the life that evolved from it about 4 billion years ago. As life evolved, the solar output increased and photosynthetic organisms used much of this carbon dioxide.

CO_2 is a major factor in the cycles that built up our 2 mineral resources. Fossil fuels developed over several hundred million years during the Phanerozoic era. There was abundant life for about 600 million years. The richest fossil fuel deposits are thought to occur at times when the earth was much warmer and contained much more CO_2 than today.

During the last 2 million years, the permanent polar ice descended and most of the evidence indicates that CO_2 levels decreased compared to the times when dinosaurs lived.

Gas bubbles found in ancient ice in Greenland and Antarctica suggest that during the end of the last great ice age (10,000 to 20,000 years), CO_2 levels were about 2/3 of what they are today. After the last ice age (5,000-9,000 years ago) the summers were about 2°F warmer than today and the CO_2 concentrations grew to preindustrial levels.

Since then, there has been a 25% increase in CO_2. The burning of organic matter may be a major part of this increase. As CO_2 goes into the atmosphere at a much higher rate than it can be withdrawn or absorbed by the oceans or living plants, there is a CO_2 buildup and this could be one of the controlling factors of the climate.

Carbon dioxide is not the only greenhouse gas that humans have been changing. Methane is another important greenhouse gas. It has increased in the atmosphere by almost 100% since 1800. Methane is produced by biological processes where bacteria have access to organic matter. These locations include marshlands, garbage dumps, landfills and rice fields. Some methane is also liberated in the process of extracting coal or transporting natural gas. Methane is 20-30 times as effective at absorbing infrared radiation as CO_2.

Methane is not as important as CO_2 in the greenhouse effect since the CO_2 percentage is much greater in the earth's atmosphere. Chlorofluorocarbons (CFCs) are even more effective greenhouse gases, but are only about 20% of the CO_2 greenhouse gases. They are involved in the destruction of stratospheric ozone.

Ozone is a form of oxygen (O_3), where three oxygen atoms combine into one molecule. Ozone has the property of absorbing most of the sun's ultraviolet radiation. It does this in the upper part of the atmosphere (the stratosphere) which is between about 10-50 kilometers (6-30 miles) above the earth. This absorption of ultraviolet energy causes the stratosphere to heat up. Life on earth has been dependent on the ozone layer shielding use from harmful solar ultraviolet radiation. Ozone is part of the earth's greenhouse effect, although it is not as important as CO_2 or methane. Ozone in the lower atmosphere can cause damage to plant or lung tissues and is a pollutant in photochemical smog.

Other trace greenhouse gases, include nitrous oxide (laughing gas), carbon tetrachloride, and several other minor gases. The collective greenhouse effect is thought to add between 50-150% to the increase in greenhouse effect expected from CO_2 alone.

Greenhouse Controls

Most agree that the greenhouse effect exists and that CO_2 has increased by some 25%. The proposed National Energy Policy Act of 1988, called for controls on industrial and agricultural emissions producing greenhouse gases. There were:

1. Regulations to ensure energy efficiency.
2. Controls on deforestation.
3. Curbs on population growth.
4. Increased funding for energy alternatives, including safer nuclear power.

Temperature Cycles

Sulfates have been found in ice cores in a regular pattern over 150,000 years of the earth's history. The cycle occurs with the change in the earth's orbit that causes the North Pole to point toward the sun when the earth is closest to it.

The ice cores also showed that calcium carbonate seems to have a cyclic pattern. This cycle occurs with the change in the tilt angle of the earth's axis relative to the plane of its orbit. The greater the tilt angle the hotter the seasonal extremes are.

Changes in the amount of carbon dioxide closely track the temperature changes inferred over the past 150,000 years. So does methane, which is 20-30 times more potent as a greenhouse gas than CO_2.

Methane is produced largely by microbial decomposition of dead organic matter in bogs, garbage dumps and swamps. The close correlations between change in methane, CO_2, and temperature implies that biological processes are involved.

It may take 500-1,000 years for the wave of warming temperatures that occur as the ice ages drop to penetrate tundra and permafrost. This warming could release methane to the air. In bog formation vast quantities of organic matter are stored during glacial periods. These processes could be an important part of the explanation of large climatic changes that have occurred in the past.

There is almost as much carbon in the atmosphere in the form of CO_2 as there is in living matter, mostly in trees. But, there is several times more carbon in the soils stored as dead organic matter (necromass). Bacteria eventually help to decompose some of this necromass into greenhouse gases such as methane, nitrous oxide (N_2O), and CO_2. The decomposition speeds up if the soil gets warmer, emitting more greenhouse gases and increasing the warming effect.

It took about a billion years for the bacteria and algae of the earth to build up the oxygen that makes our lives possible. It took another billion years for multicelled creatures to evolve into plant and animal kingdoms. A hundred million years ago, dinosaurs roamed and the average temperature was 10° to 15°C (18° to 27°F) warmer.

There was no permanent ice at the poles that can be detected from the geologic record of that period. Then, the continents drifted. The Antarctic continent isolated itself at the South Pole. India drifted northward across the equator and connected with Asia. The Tibetan plateau rose and sea level dropped about 1,000 feet. The planet cooled and permanent ice was formed.

The combination of forces that caused these changes is still debated. One fundamental question involves the stability of the climate. The climate has fluctuated between limits of plus or minus 15°C (27°F) for hundreds of millions of years. These limits are large enough to have major influence on species' extinction and evolution.

There does not seem to be much chance that the Earth would be vulnerable to a runaway greenhouse effect such as Venus where the oceans would boil away. But, climate changes as great as an ice age could be disastrous if they occurred rapidly.

Amount of Greenhouse Gases

The definite amount of greenhouse gases in the atmosphere has yet to be calculated. It depends on population, the per capita consumption of fossil fuel, deforestation, forestation activities and countermeasures that balance the extra carbon dioxide in the air. This could include planting more trees. Another factor is the use of alternative energy and conservation. Trades in fuel carbon will also have an effect. Transfers may occur from coal-rich to coal-poor nations.

Much will depend on the capital resources available to spend on energy versus other strategic commodities such as food stuffs, fertilizers and even weaponry. Based on different rates of growth in the use of fossil fuels most projections show a 1-2% annual growth rate for fossil fuels. This could produce a doubling of preindustrial CO_2.

The different greenhouse gases may have complicated interactions. Carbon dioxide can cool the stratosphere which slows the process that destroys ozone. Stratospheric cooling can also create high altitude clouds which interact with chlorofluorocarbons to destroy ozone.

Methane can be produced or destroyed in the lower atmosphere at varying rates, depending on the pollutants that are present. It also can affect chemicals that control ozone formation. All the other greenhouse gases together are assumed to be equivalent to CO_2.

Carbon Stocks

In the control of the global distribution and stocks of carbon, one process is the uptake by green plants. Since CO_2 is the basis of photosynthesis, more CO_2 in the air means faster rates of photosynthesis. Other factors are the amount of forested and planted areas, and the effects of climate change on ecosystems.

The removal of CO_2 from the atmosphere takes place through biological and chemical processes in the oceans, which can take decades or centuries. Climate change will modify the mixing processes in the ocean.

About the same amount of carbon (almost 800 billion tons) is stored in the atmosphere as is stored in living plant matter on land, mostly in trees. Animals retain a small amount of carbon about 1-2 billion tons and the amount in humans is just a small percentage of this. Bacteria have almost as much weight in carbon atoms as all the animals together and fungi have about half that amount. Dead organic matter, mostly in soils, contains about twice as much carbon as does the atmosphere.

A major potential exists for biological feedback processes to affect the amounts of carbon dioxide that might be injected into the air over the next century.

Biological Feedback

As CO_2 increases, green plants could take up more carbon dioxide into plant tissues through photosynthesis reducing slightly the buildup of CO_2. This mechanism could moderate some of the greenhouse effect. However increasing the temperature in the soils by a few degrees could increase the activity rates of bacteria that convert dead organic matter into CO_2. This would be a positive feedback loop, since warming would increase the CO_2 produced in the soils, further increasing the warming. The EPA has called this supply of soil-organic carbon a sleeping giant, with the potential for major positive feedback that could greatly advance the greenhouse effect.

There are more than a dozen biological feedback processes that could affect estimates of the temperature sensitivity to greenhouse gases due to human activities. All these operating together in unison could double the sensitivity of the climatic system to the initial effects of greenhouse gases. This would be a possible but worst case situation. The time frame over which these processes could occur is decades to a century or more.

Although all the living matter in the oceans contains only about 3 billion tons of carbon, ten thousand times that amount is dissolved in the oceans, mostly in nonliving form. The carbonate sediments in the continental crust and the ocean floor contain almost 70 million billion tons of carbon. These are enormous quantities compared to the atmosphere and living and dead biota.

The quicker the climate warms up, the more likely it is that feedback processes will change the greenhouse gas buildup. There is a widespread agreement that CO_2 and other trace greenhouse gases may double sometime within the next century.

Climate Modeling

Most climate models provide a climate in stable equilibrium when they are forced by a doubling of CO_2. If the 1900 condition of about 300 parts per million doubles to 600 parts per million, most three-dimensional models show that an equilibrium occurs eventually at an average surface temperature warming of 3.5° to 5°C (5.6° to 9°F). If the carbon dioxide content of the atmosphere doubled in one month, the earth's temperature would not reach its new equilibrium value for a century or more.

If suddenly we were able to control all CO_2 emissions, we would still expect about one degree of warming while the climatic system catches up with the greenhouse gases already released.

It is not the global average temperature that is most important but it is the regional patterns of climate change. Making reliable predictions of local or regional responses requires models of great complexity, but most calculations suggest:

- wetter subtropical monsoonal rain belts
- longer growing seasons in high latitudes
- wetter springtimes in high and middle latitudes
- drier midsummer conditions in some midlatitude areas
- increased probability of extreme heat waves
- increased probability of summertime fires in drier/hotter regions
- increased sea levels by a meter or more (several feet) over the next hundred years.

There are health consequences for people and animals in already warm climates with a reduced probability of extreme cold snaps in colder regions.

Countermeasures would include steps to prevent atmospheric changes through engineering modifications. Strict economists favor doing nothing active now, assuming that resources will be used to maximize economic conditions in the future and solutions will eventually develop. Strict environmentalists favor a redistribution of resources to modify costs and incomes.

Moderating the Greenhouse Effect

A number of proposals have been made for reducing the amount of CO_2 in the free atmosphere as a means of moderating the greenhouse effect. Prescrubbing involves taking the carbon out of the fuels prior to combustion, leaving only the hydrogen to be burned. Another technique is postcombustion scrubbing which removes CO_2 from the emissions stream after burning but before release to the atmosphere.

One prescrubbing technique is the hydrocarb process, where hydrogen is extracted from coal and the carbon is then stored for possible future use or buried. In this process, only about 15% of the energy in coal would be converted to hydrogen for use as fuel in existing coal power plants and there would be much residual solid coal material to store. Also, the hydrogen generated by this process would have to be transported. The existing pipeline system is currently limited to an annual flow of about 25 trillion cubic feet of gas in the United States which is insufficient to operate the present utility structure. If natural gas is used, there is also the question of pipeline capacity.

In a report to the Department of Energy, Pacific Northwest Laboratory estimated capital costs to be several trillion dollars. This is about $8,000 per capita in the United States for 300,000 megawatts of generating capacity to replace the coal consumed in the U.S. for electrical power generation. Post combustion scrubbing was called a known but unapplied technology. Removing 90% of the CO_2 from the stack gases would cost about 0.5 to 1 trillion dollars or $2,000 to $4,000 per capita. Removing the CO_2 at a power plant could use up about half the energy output of the power plant.

Pumping the carbon dioxide produced at industrial plants into the deep ocean was proposed in 1977 by Cesare Merchetti of the International Institute of Applied Systems Analysis (IIASA) near Vienna, Austria. This would reduce and delay the rise of carbon dioxide in the atmosphere, but would not prevent an eventual warming as some made its way back into the atmosphere.

Using reforestation as a carbon bank would capture carbon from the atmosphere, but the decay or burning of the harvested trees decades later would have to be prevented. Burial in mines could prevent the release of the CO_2 into the atmosphere.

Vegetational carbon banks would compete with agriculture for land and nutrient resources. It is estimated that a land area about the size of Alaska would need to be planted with fast-growing trees over the next 50 years to use up about half the projected fossil-fuel-induced CO_2 at a cost of about $250 billion or $50 per person for the global population. One problem is that once the trees are fully grown they no longer take up CO_2 very rapidly and would need to be cleared so new trees could be planted to continue a quicker uptake. Old trees could be used for lumber, but not fuel, since this would release the CO_2. If used as fuel, a delay of 50 years, (the typical growth time) would occur and move up the buildup rate of atmospheric CO_2.

Proposals for counteracting global warming from the greenhouse effect include releasing dust or other particles to reflect away part of the solar energy normally absorbed by earth. This could work on a global average basis, but the mechanisms of warming and cooling would vary and large regional climatic changes could still occur.

Energy Conservation

Energy conservation could help reduce the impact of many problems. Increasing energy efficiency could reduce atmosphere pollution on most fronts while improving national security through increased energy independence.

The environmental effects of carbon dioxide and acid rain would be reduced along with the risk of possible climatic changes. The United States uses twice as much energy in manufacturing than Japan, West Germany, or Italy. The cost of this energy keeps the cost of products in the U.S. higher.

Developing nonfossil energy sources and improving efficiency in all energy sectors should be viewed as part of a high-priority strategic investment. The mechanisms to accomplish this include research and development on solar photovoltaic cells and safer nuclear plants and possible tax incentives to reduce fossil fuel emissions.

Greenhouse gas buildup is a global problem and is connected to global economic development. It depends on population, resources, environment and economics. Developed countries are the major producers of CO_2.

Global strategies for preventing CO_2 buildup requires international cooperation between rich and poor nations. The increased burden of debt is a major hurdle in the global development of the Third World. It is difficult for countries to invest in expensive energy-efficient equipment when they can hardly pay back the interest on loans from other countries. A debt/nature swap has been proposed where underdeveloped countries would provide tracts of forest to developed countries in exchange for forgiving part of their debt. The United States is now the world's biggest debtor. Another approach is to have the World Bank place environmental conditions on its loans.

Population growth rates are another point of contention between developed and developing countries. Total emission is the per capita emission rate times the total population size. The population growth which is occurring predominantly in the Third World will become an important factor.

For the poor of the world, more energy and more energy services can mean a better quality of life. Energy use can allow services that improve health care, education and nutrition in less-developed nations.

As population or affluence grows, so does pollution. Ultimately the world population should stabilize and future pollution levels should be lower for any per capita standard of consumption. A stable population is an essential part of a sustainable future. There is the risk in increasing pollution now in order to achieve the improved quality of life necessary for a population size that is sustainable.

If the buildup of CO_2 and other trace gases is not considered as part of global development, it is unlikely that greater buildup will be prevented, except by great advances in alternative fuel systems and programs to increase energy efficiency.

We produce about 1.2 billion metric tons of carbon each year in the form of carbon dioxide. We would have to reduce this to approximately 1 billion metric tons per year.

Proposed legislation in 1988 called for a 50% reduction in CO_2 in the United States early in the next century which equates to reducing emissions to about 650 million metric tons per year. Our residential and commercial energy use is about 15% of total energy used. About half of this involves natural gas. Industrial energy use is about 20%, almost evenly divided between oil and gas with a substantial minority depending on coal. Transportation is also about 20%, most of which is derived from oil. Electric utilities account for 35%, most of this is produced by coal, with half that amount again from nuclear, and about half that amount from natural gas. This accounts for most of the energy use in the United States.

Since coal is the least efficient fuel, it produces the greatest amount of CO_2 per unit energy. Any growth in applications that use coal would substantially increase CO_2 levels. Moving to natural gas, nuclear, solar, hydro or wind power would decrease CO_2 amounts.

Hydrogen could become a major energy source, reducing U.S. dependence on imported petroleum while diversifying energy sources and reducing pollution and greenhouse gas emissions. It could be produced in large refineries in industrial areas, power parks and fueling stations in communities, distributed facilities in rural areas with processes using fossil fuels, biomass, or water as feedstocks and release little or none carbon dioxide into the atmosphere.

By 2020 hydrogen could be used in refrigerator-sized fuel cells to produce electricity and heat for the home. Vehicles that operate by burning hydrogen or by employing hydrogen fuel cells will become commercially viable and emit essentially water vapor. Hydrogen refueling stations using natural gas to produce hydrogen may be available in urban areas to refuel hydrogen vehicles.

Micro-fuel cells using small tanks of hydrogen could be operating mobile generators, electric bicycles and other portable items. Large 250-kW stationary fuel cells, alone or in tandem, are being used for backup power and as a source of distributed generation supplying electricity to the utility grid.

There are several benefits to be expected from a hydrogen economy. The expanded use of hydrogen as an energy source should help to address concerns over energy security, climate change and air quality.

Hydrogen can provide a variety of domestically produced primary sources including fossil fuels, renewable, and nuclear power and allow a reduction of the dependence on foreign sources of energy. The by-products of hydrogen conversion are generally agreeable to human health and the environment.

One Department of Energy study compared alternative paths for future U.S. energy use: business-as-usual and energy-efficient. Both projections suggested a substantial rise in U.S. production of CO_2 and the consumption of fossil fuels over the next several decades. This study was the DOE's National Energy Policy Plan (NEPP) projection to 2010. It was prepared in 1985 and was a study where environmental effects were the primary objective. The study predicted an increase in energy between 1985 and 2010 of about 30% while the projected oil and gas consumption remained relatively constant over this period. However, coal consumption increased greatly by more than 100%. CO_2 emissions increased from 1.25 billion metric tons per year in 1985 to about 1.73 billion metric tons in 2010. This is a 38% increase in CO_2.

During 1975 to 1985 dramatic gains in energy efficiency in the United States lowered fossil fuel emissions while the gross national product increased. The gains in energy efficiency were driven by the OPEC oil price jumps.

However the 1975-1985 period was one in which economic growth and energy growth remained relatively unlinked. Most historic periods show the reverse trend. The DOE report views this period as a deviation.

The main reason for the 38% increase in CO_2 was a more than doubling of the coal use in electric utilities and a near doubling of coal use in industrial use. Coal was the fifth largest U.S. export in 1982 and there would be some pressure to keep this source of income, given our large trade deficit.

The energy efficiency path still increased CO_2 production to 1.5 billion metric tons per year, almost 230 million tons more than 1985 emissions. The high-efficiency case does use less coal. Other energy analysts predict a decline in energy-growth rates and a decline instead of an increase in CO_2 emissions.

The DOE forecast views U.S. energy production under a freer economic viability and stronger technological growth for energy systems.

Others see an energy future tied to broad societal goals of economic efficiency and equity and assumes policy changes can achieve the objectives. Energy service demands are developed based on assumed levels of per capita income. Market interventions are promoted which would reduce the energy supply. These could include petroleum product taxes and oil import fees and carbon taxes for greenhouse problems.

The shift in the U.S. economy from energy-intensive activities such as steel manufacturing to information-intensive activities such as computer and software design will continue to improve our gross national product while reducing our dependence on oil and coal.

Many studies assume improvements in the gas mileage of cars and efficiency in the production of energy in power plants, in industrial applications and in home heating, lighting and other sectors. U.S. manufacturers could improve the average energy efficiency of cars and trucks. But, as America's fleet of older vehicles is replaced with newer cars with less pollution, CO_2 emissions may change very little or even increase since additional miles may be driven.

The NEPP high-efficiency case assumed that by the year 2010 new cars would average 52 miles per gallon and would penetrate 50% of the U.S. market. This would be possible if small hybrids take over a major part of the market.

Other studies believe that new car efficiencies could be even greater than that, with a fuel economy for the average vehicle of 75 miles per gallon.

The NEPP report also assumed that the U.S. economy would reduce its dependence on energy at the rate of about 1.7% per year, while others believe that more active efforts to make our economy less dependent on energy could result in a rate of about 4% per year. They also assume very high-efficiency lighting and the rapid deployment of electric-power-generating stations that are 50% or more efficient than present facilities.

Improvements in efficiency along with major efforts to redirect energy use towards greater environmental quality and source reliability would not only reduce emissions but there would be many other benefits.

In 1950 the U.S. CO_2 emissions were almost 40% of the global total. By 1975 this had dropped to about 25%, and by the late 1980s it was about 22%. If the U.S. held emissions constant at 1985 levels, a reduction of 15% from the emissions in 1995 and a 28% reduction from the forecast emissions in 2010, then global emissions would be reduced by only 3% in 1995 and 6% in 2010. Even if U.S. emissions were reduced by 50% below the 1985 levels, global emissions would continue to grow and would be cut by less than 15% in the year 2010. This proves the assumption that world emissions will continue to grow.

Indirect effects are also possible, since it is likely that if the United States introduced technological improvements to reduce CO_2 emissions, then the resulting cost reductions would provide a competitive advantage and would be imitated by foreign competitors. This response would then reinforce U.S. emission reductions, leading towards a worldwide effect. We could also develop crop strains that could take advantage of CO_2. Some climate changes cannot be prevented, but it may be possible to minimize any damages from an altered climate.

Preventive Strategies

Certain preventive strategies could actively limit emissions of substances thought to be harmful. One strategy designed to avoid damage to the ozone layer was the reduction or banning of all uses of CFCs. The Montreal Protocol of 1987 proposed a 50% cut in CFCs by the year 2000, but not all nations signed the treaty. Most scientific studies push for at least a 90% ban if the ozone hole is to be reversed. This would not only help protect the ozone layer but would cut emissions of a trace greenhouse gas that could be responsible for up to 25% of global warming.

The use of artificial fertilizers in agriculture also generates atmospheric nitrogen compounds that can reach the stratosphere and possible destroy ozone.

The present theories of the origin of acid rain suggest that we can limit acid rain by reducing sulfur dioxide emissions and switching to low-sulfur fuels. Only about 20% of the world's petroleum reserves are low in sulfur. Switching U.S. Midwestern power plants to low-sulfur coal could cause economic problems since much of the coal from the Midwest and Appalachia has a high sulfur content.

The electric power industry in the Midwest, that uses much of the high-sulfur coal, would have to spend tens of billions of dollars to scrub the sulfur out of coal. An energy penalty would also be paid for the processes that remove the sulfur along with environmental problems from disposing of it. Approximately 5% more coal would have to be burned to keep electricity production from these power plants at current levels if most of the sulfur is scrubbed out.

It is possible to keep sulfur dioxide from reaching the atmosphere by washing the coal or by removing the SO_2 from the flue gas. Simple washing can remove about 50% of the sulfur. Additional removal of up to 90% requires high temperatures and high pressures and can cost ten times as much as washing. Flue gas desulfurization (scrubbing) by reacting the effluent gas with lime or limestone in water can remove 80-90% of the sulfur but creates large amounts of solid waste.

Techniques for minimizing emission of SO_2 from burning power plants have no effect on nitrogen oxide (NO_x) emissions. Oxides of nitrogen result from the burning of nitrogen normally found in combustion air. The percentage of NO_x generated by the burning of air is about 80% in conventional coal-fired boilers and depends mostly on the temperature of combustion. Improved furnace designs and combustion techniques could reduce NO_x emissions from stationary sources by 40-70%. These methods are not in widespread use now. The processes for removing NO_x from flue gases are in an early stage of development.

Replacing old inefficient electricity-generating plants with vastly more efficient new plants would save much energy. Some of these older plants lose two-thirds of the heat energy as waste heat at the site. Replacing these plants with those with more efficient boilers, controls and turbines would reduce the lose of heat energy to about half. The plants could also switch from coal to natural gas which would dramatically reduce the acid rain problem and cut CO_2 emissions in half.

Emissions from automobiles could be decreased with improvements in the design of combustion chambers and the computer control of combustion mixtures. Exhaust-gas catalytic converters can also limit emissions.

Battery and fuel cell-powered cars could dramatically reduce air pollution in cities, but only if the electric power sources used to charge the batteries or create the hydrogen fuel were themselves less polluting. Energy-efficient mass transit can also reduce mobile-source emissions of NO_x as well as CO_2.

Coalition for Vehicle Choice

The coalition for Vehicle Choice (CVC), which is a lobbying group sponsored by carmakers, has campaigned to repeal the CAFE standards. The CVC has claimed that CAFE causes 2,000 deaths and 20,000 injuries every year by forcing people into smaller cars. The auto industry has questioned the science behind global warming and claimed there is not enough information to make a judgment.

Toyota was the first auto company to announce, in the spring of 1998, that it was joining others such as British Petroleum, Enron, United Technologies, and Lockheed Martin in an alliance to fight global warming. Toyota is supporting the Pew Center on Global Climate Change, which was funded with a $5 million start-up grant from the Pew Charitable Trusts.

Global Warming

Carbon dioxide, in combination with other greenhouse gases such as methane and ozone, can trap the sun's heat. In the century from 1890 to 1990, the average surface temperature of the earth increased by 0.3 to 0.6 degrees Celsius. This temperature rise, which has lengthened the growing season in parts of the northern hemisphere, may have occurred naturally, although such a change is unheard of in the last 600 years of recorded history.

The global warming issue has become a national affair. It was not that long ago that global cooling occupied our attention. In the 1970s, several extreme weather events, including freezing conditions in Florida, produced fears over temperature decline. In 1974, the CIA even issued a report stating that decreases in temperature could affect America's geopolitical future.

During the 1980s, the national focus shifted to global warming, as a result of the unusual drought and heat wave of 1988. Climate scientist James Hansen stated to Congress that he was 99% sure that the greenhouse effect was contributing to global warming. Immediately, a growing anxiety over rising temperatures had the attention of the media and the public.

Environmentalists seized on the human role in global warming to advance goals such as improving air quality and preserving forestland. Some studies indicate that human influence accounts for 75% of the increase in average global temperature over the last century. But, scientists are not in agreement on what accounts for the warming trend. Changes in global ocean currents or in the amount of energy emitted by the sun could be a major part of the change.

Most of industry including the oil, gas, coal and auto companies view the problem as a theory in need of more research. Many studies call for serious action to reduce fossil fuel use.

Many scientists do not debate whether global warming has occurred, they accept it. But, the cause of the warming and future projections about how much the earth will heat up divides many.

In the early 1990s, the Information Council on the Environment, which was a group of coal and utility companies, used a public relations firm to advance global warming as theory. The U.S. auto industry has also played a role. Lobbying was done to dismiss global climate change and fight legislation on fuel economy which is an important factor in carbon emissions.

Fuel Economy Standards

In the 1980s, U.S. carmakers succeeded in getting the government to relax fuel economy standards, by arguing that they would have to close down factories to meet stricter requirements. In 1973, when the energy crisis began, American automobiles averaged about 13 miles per gallon of gasoline (MPG). By the early 1990s, that number had increased to almost 27 MPG. This seems like a significant increase, but its represent less than one mile to the gallon improvement per year.

The number of vehicle miles traveled doubled between 1970 and 1990 to 2.2 trillion miles, so these gains in fuel efficiency were mostly offset. The improvement in efficiency was mainly due to legislation passed in 1975 that established Corporate Average Fuel Economy or CAFE rules. This allowed automakers to produce any kind of car as long as all the vehicles when averaged meet the MPG standards set by the government.

In 1992, Bill Clinton campaigned for president on the promise that he would increase the CAFE standard to 45 MPG. During this same period, President Bush signed a global warming treaty at the 1992 Earth Summit in Rio de Janeiro, Brazil. Under the treaty, industrialized nations agreed by the year 2000 to voluntarily cut back their carbon dioxide emissions to the level they were at in 1990. To meet this goal, U.S. vehicles would need to be three to four times more efficient than they were, averaging about 80 to 90 MPG.

The auto industry balked at Clinton's 45 MPG goal. When elected, Clinton broke his promise. In 1993, the administration announced that the federal government would join up with American automakers to produce a new, super-efficient car.

Clean Car Initiative

The clean car initiative diverted attention from one of the most important developments in automotive history. In the last few decades the sport utility vehicle or SUV has become dominant. Many have attacked SUVs and light trucks as vehicles that use too much gas and cause excess pollution compared to the typical sedan.

Light trucks were excluded from the 1975 fuel economy legislation. It was argued that farmers and construction workers used them for business purposes. Actually, many trucks are used in the same capacity as cars. The Chrysler minivan also fit into this category.

By the 1990s, U.S. carbon emissions were rising. Americans were spending more time on the road and traveling in more of the least fuel-efficient vehicles. Minivans, SUVs, and pickup trucks make up about 40% of all vehicles sold in the United States. By 1999 SUVs were getting larger and larger. Some are more than 18 feet long and weigh as much as 12,500 pounds which is about as much as four mid-sized sedans. Fuel economy is about 10 miles per gallon.

The average fuel economy of new cars and trucks sold in the United States during the 2001 model year remains at its lowest level in two decades.

In 1997, the parties of the earlier Rio treaty went to Kyoto, Japan to work on carbon emission standards for the industrialized nations. Later that year, the U.S. Senate voted 95-0 against ratifying any global warming treaty that did not require developing nations themselves to reduce carbon emissions.

In 2000 at the Netherlands, global warming talks broke down over carbon accounting. The United States wished to use its forest area to offsetting some carbon emissions.

Carbon Rights

This trading of carbon rights was the type of approach that most mainstream environmental groups in the United States had advocated in an attempt to give business an incentive to conserve. In Europe, environmentalists have taken an uncompromising stand against industry and it is looked at as a plan for evading responsibility for cleaning up the global atmosphere.

Deforestation

Growing fossil fuel use in the 20th century changed the carbon history of the earth. But deforestation also had an impact on carbon in the atmosphere. Forests serve as carbon sinks, producing oxygen while using carbon dioxide.

The clearing of forests in the United States early in the century, combined with a large increase in postwar tropical deforestation, where much of the wood was burned, releasing carbon dioxide to the air has reshaped atmospheric components.

In 1900, carbon dioxide levels were about 300 parts per million (ppm). By the 1950s, these levels had increased to 320 ppm and by the 90s to 360 ppm.

References

Carless, Jennifer, *Renewable Energy*, Walker and Company, New York,: 1993.

Cothran, Helen, Book Editor, *Global Resources: Opposing Viewpoints*, Greenhaven Press,: San Diego, CA, 2003.

Romm, Joseph J., *The Hype About Hydrogen*, Island Press: Washington, Covelo, London, 2004.

CHAPTER 7

HYDROGEN SOURCES, BIOMASS AND WIND POWER

Most of the hydrogen used in the chemical and petroleum industry is manufactured from natural gas, which is a hydrocarbon molecule of four hydrogen atoms bonded to one carbon atom. Gasoline is a hydrocarbon molecule that is made up of eighteen hydrogen atoms that are attached to a chain of eight carbon atoms. High temperature steam is used to separate the hydrogen from the carbon. If the cost of the natural gas is $4 per million British thermal units (MMBtu), the cost of the gaseous hydrogen will be about $10.00 per MMBtu. If the hydrogen is liquefied, an additional $8.00 to $10.00 per MMBtu must be added to the cost of the gaseous hydrogen, making the cost of liquid hydrogen produced by this method about $20.00/MMBtu. If hydrogen is manufactured from water with electrolysis equipment, its cost is roughly equivalent to $5/MMBtu per 10 mills ($5/kWh/cent/kWh). Table 7-1 illustrates the major sources of global hydrogen production.

Table 7-1. Global Hydrogen Production

Origin	Percent
Natural gas	48
Oil	30
Coal	18
Electrolysis	4

Hydrogen can also be manufactured from coal-gasification facilities at a cost that ranges from $8 to $12 per MMBtu, depending on the cost of coal and the method used to gasify it. But, making hydrogen from nonrenewable fossil fuels does not solve the problem of diminishing resources or the environmental problems.

Most of the easy-to-get oil has already been found, and increasingly, exploration efforts have to drill in areas that are more difficult. Many areas have been closed to drilling in the United States. At some point, in the future, it may take more energy to extract the remaining fossil fuels than the energy they contain.

Hydrogen can also be produced from resources that are renewable, such as the direct and indirect sources of solar energy, this includes the large quantities of agricultural wastes, sewage, paper and other biomass materials that have been accumulating in landfills.

Generating hydrogen from such waste materials may turn out to be one of the least expensive methods of producing hydrogen since this resource is quite extensive. It has been estimated that in the U.S., roughly 14 quads of the annual 64 quad total energy requirement could be met from renewable biomass sources, which is about 20% of our total energy needs.

Sewage in vast quantities of billions of gallons per day could be recycled to produce a renewable source of hydrogen. This can be accomplished either by utilizing the non-photosynthetic bacteria that live in the digestive tracts and wastes of humans and other animals, or by pyrolysis-gasification methods. Advanced sewage treatment systems could turn the billions of gallons of raw sewage that is being dumped into rivers and oceans into relatively low-cost hydrogen.

Although high-temperature nuclear-fusion reactors may some day be practical as renewable sources of energy for hydrogen production, they are probably many years away. Typically, over 100 million °F temperatures are required for nuclear fusion to occur and this technology, while under development, is not expected to be commercially viable in the near future.

HYDROGEN FROM NATURAL GAS

Natural gas is the least expensive source of hydrogen today. But, there may not be enough natural gas to meet the demand for natural gas power plants and to supply a hydrogen fueled economy. The prices of natural gas, hydrogen and electricity could see dramatic increases as the demand for natural gas to make hydrogen increases.

The delivered cost of hydrogen from natural gas would need to become competitive with the delivered cost of gasoline. The infrastructure costs must be managed over time with estimates reaching a trillion dollars or more.

It is not known which will be cheaper and more practical, electrolysis or reforming methane at small local filling stations or at large centralized plants. Technological advances are sure to change many aspects and questions.

Water Use

Water is a common source of hydrogen. Electrolysis is the process of decomposing water into hydrogen and oxygen using electricity. It is a mature technology widely used around the world to generate very pure hydrogen. But, it is energy intensive and the faster you generate the hydrogen, the more power that is needed per kilogram produced.

Commercial electrolysis units require almost 50-kWh per kilogram, which represents an energy efficiency of 70%. This means that more than 1.4 units of energy must be provided to generate 1 energy unit in the hydrogen.

Most electricity comes from fossil fuels, and the average fossil fuel plant is about 30% efficient, then the overall system efficiency is close to 20% (70% times 30%). Five units of energy are needed for every unit of hydrogen energy produced.

Forecourt Plants

Larger electrolysis plants are cheaper to build (per unit output) and they would pay a lower price for electricity than smaller ones at local filling stations, which are sometimes called forecourt plants since they are based where the hydrogen is needed.

Hydrogen can be generated at off-peak rates, but that is easier to do at a centralized product facility than at a local filling station, which must be responsive to customers who typically do most of their fueling during the day and early evening, the peak power demand times.

To circumvent peak power rates, the National Renewable Energy Laboratory (NREL) suggests that forecourt plants would need large over-sized units operated at low utilization rates with large amounts of storage. This requires additional capital investment. The estimated cost of producing and delivering hydrogen from a central electrolysis plant is $7-$9/kg. The cost of production at a forecourt plant may be $12/kg. High cost is probably the major reason why only a small percentage of the world's current hydrogen production comes from electrolysis. To replace all the gasoline sold in the United States today with hydrogen from electrolysis would require more electricity than is sold in the United States at the present time.

From the perspective of global warming, electrolysis is questionable in the foreseeable future because both electrolysis and central-station power generation are relatively inefficient processes, and most U.S. electricity is generated by the burning of fossil fuels. Burning a gallon of gasoline releases about 20 pounds of CO_2. Producing 1-kg of hydrogen by electrolysis would generate on average, 70 pounds of CO_2. A gallon of gasoline and a kilogram of hydrogen have about the same energy, and even allowing for the potential doubled efficiency of fuel cell vehicles, producing hydrogen from electrolysis could make global warming worse. These economic and environmental questions may make it difficult to pursue generation of significant quantities of hydrogen from the U.S. electric grid in the near future.

Hydrogen could be generated from renewable electricity, but the renewable system most suitable for local generation, solar photovoltaics, makes hydrogen that is expensive. The least expensive form of renewable energy, wind power, is only a few tenths of 1% of all U.S. generation, although that figure is rising rapidly it still has a long way to go.

Generating hydrogen from electrolysis powered by renewables is viewed by some as a good use of that power for economic and environmental reasons. But, would the United States need excess low-cost renewable generation before it can divert a substantial fraction to the production of hydrogen?

If forecourt hydrogen generation from solar photovoltaics becomes practical in the first half of the century, it could supply enough hydrogen to the growing amount of fuel cell cars and generating systems.

If hydrogen is generated from the vast wind resources in the Midwest, there would be large infrastructure costs for delivering it to other parts of the country.

FUEL CELL GENERATORS

Fuel cells can be used to generate electricity, heat, and hydrogen. FuelCell Energy uses this technique in its molten carbonate fuel cell. Some solid oxide fuel cell (SOFC) companies are developing similar products.

Fuel cells running on natural gas typically use about three of the four hydrogen atoms in methane (CH_4) for power generation. The remaining hydrogen goes into the flue gas or stack effluent with differing amounts of CO_2, CO, and water vapor, depending on the type of fuel cell. The flue gas is sometimes vented to the atmosphere but it can be combusted for heat and used for reforming.

Hydrogen can be separated from the flue gas at low cost in high-temperature fuel cells. A SOFC system may be able to cogenerate hydrogen for about $3.00 per kg. This may be a little more expensive than gasoline or it might even match gasoline. Since these fuel cells could be part of the fueling station, there would be no need for a hydrogen delivery infrastructure. This requires fuel cells to achieve important technical and economic goals and overcome the barriers to utilization.

Coal is another source of hydrogen. The coal is gasified and the impurities are removed so the hydrogen can be recovered. This results in significant emissions of CO_2.

GASIFICATION TECHNOLOGIES

The Tampa Electric Company plant in Polk County, Florida uses coal gasification to generate some of the nation's cleanest electricity. Coal gasification represents the next generation of coal-based energy production. The first pioneering coal gasification power plants are now operating in the United States and other nations. Coal gasification is gaining increasing acceptance as a way to generate extremely clean electricity and other high-value energy products.

Instead of burning coal directly, coal gasification reacts coal with steam and carefully controlled amounts of air or oxygen under high temperatures and pressures.

The heat and pressure breaks the chemical bonds in coal's complex molecular structure with the steam and oxygen forming a gaseous mixture of hydrogen and carbon monoxide. Gasification may be one of the better ways to produce hydrogen.

Pollutants and greenhouse gases can be separated from the gaseous stream. As much as 99% of sulfur and other pollutants can be removed and processed into commercial products such as chemicals and fertilizers. Unreacted solids can be collected and marketed as co-products such as slag for road building.

The primary product is fuel-grade, coal-derived gas which is similar to natural gas. The basic gasification process can also be applied to other carbon-based feedstocks such as biomass or municipal waste.

Coal gasification offers a more efficient way to generate electric power than conventional coal-burning power plants. In a conventional plant, heat from the coal furnaces is used to boil water, creating steam for a steam-turbine generator.

In a gasification-based power plant, the hot, high pressure coal gases from the gasifier turn a gas turbine. Hot exhaust from the gas turbine is then fed into a conventional steam turbine, producing a second source of power. This dual, or combined cycle arrangement of turbines is not possible with conventional coal combustion. It offers major improvements in power plant efficiencies.

Conventional combustion plants are about 35% efficient (fuel-to-electricity). Coal gasification could boost efficiencies to 50% in the near term and to 60% with technology improvements. Higher efficiencies mean better economics and reduced greenhouse gases.

Compared to conventional combustion, carbon dioxide exits a coal gasifier in a concentrated stream instead of a diluted flue gas. This allows the carbon dioxide to be captured more easily and used for commercial purposes or sequestered.

Historically, the use of gasification has been to produce fuels, chemicals and fertilizers in refineries and chemical plants. DOE's Clean Coal Technology Program allowed utilities to build and operate two coal gasification power plants; Tampa, Florida and West Terre Haute, Indiana. A Clean Coal Technology gasification project is also operating at Kingsport, Tennessee producing coal gas that is chemically recombined into industrial grade methanol and other chemicals. Gasification power plants are estimated to cost about $1200 per kilowatt, compared to conventional coal plants at around $900 per kilowatt.

The Vision 21 program is focused on new concepts for coal-based energy production where modular plants could be configured to produce a variety of fuels and chemicals depending on market needs with virtually no environmental impact outside the plant's footprint. Membranes would be used to separate oxygen from air for the gasification process and to separate hydrogen and carbon dioxide from coal gas.

Improved gasifier designs would be more durable and capable of handling a variety of carbon-based feedstocks. Advanced gas cleaning technologies would capture virtually all of the ash particles, sulfur, nitrogen, alkali, chlorine and hazardous air pollutants.

The Clean Coal Power Initiative would spend $2 billion over the next 10 years for these high-potential, but still high-risk, technologies. Targets are an efficiency greater than 52% with emissions of NO_x 0.06, lb/million Btu and SO_2 0.06 lb/million Btu and a cost of less than $1,000/kW by 2008.

Estimates for producing and delivering coal-generated hydrogen range from $4.50 to $5.60/kg, which is more than the cost of U.S. gasoline on an equivalent energy basis.

Coal is the most abundant fossil fuel in the U.S. and many other countries. In the U.S. coal makes up about 95% of all fossil energy reserves. These reserves could last several hundred years at the current level of coal consumption. Major developing countries such as China and India, which are now using more and more of the world's oil, also have large coal reserves.

Many countries and companies have channeled R&D efforts into generating hydrogen and electricity from coal without releasing CO_2. Gasification and cleaning can be used that combines coal, oxygen or air, and steam under high temperature and pressure. The process generates a synthesis gas (syngas) of hydrogen and CO_2. The syngas does not contain impurities such as sulfur or mercury. A water-gas shift reaction is then used to increase hydrogen production and create a stream of CO_2 that can be removed and piped to a sequestration site. The hydrogen-rich gas is sent to a Polybed Pressure Swing (PSA) system for purification and transport. The remaining gas that comes out of the PSA system can be compressed and sent to a combined cycle power plant. These are similar to the natural gas combined cycle plants used today.

Hydrogen as well as syngas may also be used to power a combined cycle plant. The plant output can be adjusted to generate more power or more hydrogen as needed.

Cogeneration of hydrogen and electricity from coal, coupled with CO_2 extraction needs to be an affordable and practical system for generating both energy carriers. Hydrogen could be generated from large coal plants outside cities, close to existing coal mines, then the infrastructure costs for delivering the hydrogen could be high.

FutureGen

In 2003 the Department of Energy announced FutureGen. This is also known as the Integrated Sequestration and Hydrogen Research Initiative.

This is a 10-year, billion dollar project to produce a 275-MW prototype plant that will cogenerate electricity and hydrogen and sequester 90% of the CO_2. This advanced coal-based, near-zero emission plant is planned to produce electricity that is only 10% more costly than current coal-generated electricity while providing hydrogen that can compete with gasoline. The cost of hydrogen delivery is not included in this goal.

A 2002 study for the National Energy Technology Laboratory found that coal gasification systems with CO_2 capture could reach efficiencies of 60% or more in cogenerating hydrogen and electricity using different configurations of turbines and solid oxide fuel cells (SOFCs).

Building large commercial coal gasification combined cycle units could be difficult based on the history traditional power generators have had with simpler chemical processes. Sequesting the CO_2 will be another technological challenge.

Sequestration

Sequestration involves storing CO_2 in large underground formations. CO_2 separation and capture are part of many industrial processes, but using existing technologies would not be cost-effective for large-scale operations. Sequestration costs using current technology are high.

The practicality and environmental consequences of many sequestration techniques have not yet proven from an engineering or scientific aspect. Sequestration still requires much research and development before generating large volumes of hydrogen from coal and sequestering the CO_2 produced. CO_2 sequestration on a massive scale would need to be permanent to be practical.

HYDROCARBON SOURCES

Gasoline may be used as a source of hydrogen. Hydrogen can be produced from hydrocarbons such as gasoline and methane using partial oxidation and autothermal reformers. High cost is one problem for onboard gasoline reformers, the other is the high temperature at which they operate does not allow a rapid starting. The reforming process also involves a loss of about 20% of the energy in the gasoline.

In 2003, Nuvera Fuel Cells developed a 75-kW gasoline reformer. It has better than 80% efficiency but takes over a minute to start. Nuvera has been working to get this down to 30 seconds and thinks the reformer would eventually sell for about $2,000.

Gasoline fuel cell vehicles (FCVs) may be an interim step. Their main advantage is the use of an available fuel. An affordable gasoline reformer could allow a market for fuel cell vehicles without a hydrogen infrastructure.

Methanol or wood alcohol is another potential source or carrier of hydrogen. Methanol, CH_3OH, is a clear liquid, the simplest of the alcohols, with one carbon atom per molecule. Methanol is extensively used today, the U.S. demand in 2002 was over a billion gallons. The largest U.S. methanol markets were for producing the gasoline additive MTBE (methyl tertiary butyl ether) as well as formaldehyde and acetic acid. MTBE is being phased out since it has been found to contaminate water supplies.

Methanol is already used as an auto fuel. It has been the fuel of choice at the Indianapolis 500 for more than three decades, partly because it improves the performance of the cars but it is also considered much safer. It is less flammable than gasoline. When it does ignite, it causes less severe fires. A study for the U.S. Environmental Protection Agency (EPA) concluded, that the use of methanol can result a 90% reduction in the number of automotive fuel related fires compared to gasoline.

Methanol also seems to biodegrade quickly when spilled and it dissolves and dilutes rapidly in water. It has been recommended as an alternative fuel by the EPA and the DOE, partly because of reduced urban air pollutant emissions compared to gasoline. Most methanol-fueled vehicles use a blend of 85% methanol and 15% gasoline called M85.

Methanol has several advantages for powering fuel cell vehicles. A study for the CAFCP pointed out methanol's availability without new infrastructure, high hydrogen-carrying capacity and the ability to be stored, delivered, and carried onboard without pressurization.

Our present transportation system and its infrastructure favor liquid fuels. Fuel cell vehicles with onboard methanol reformers would have very low emissions of urban air pollutants. Daimler-Chrysler has built demonstration fuel cell vehicles that convert methanol to hydrogen.

Methanol reformers operate at lower temperatures (250°C-350°C), so they are more practical than onboard gasoline reformers. Methanol reformers could also be used at fueling stations to generate forecourt hydrogen.

Direct methanol fuel cells (DMFCs) could run on methanol without a reformer. Practical, affordable DMFCs for cars and trucks appear to be years away.

Methanol is mainly synthesized from natural gas, it can also be produced from a number of CO_2-free sources, including municipal solid waste and plant matter.

It is also toxic, as the EPA has noted, and a few teaspoons of methanol consumed orally can cause blindness. A few tablespoons can be fatal, if not treated. Methanol is also very corrosive, so it requires a special fuel-handling system.

Most major oil companies that are involved in delivering our transportation fuels are not enthusiastic about a methanol economy. Methanol has been used to make MTBE, a gasoline additive now being phased out in California because of environmental concerns such as groundwater contamination. Although methanol exists in nature and degrades quickly, MTBE is a complex, compound that exhibits little degradation once released into the environment.

For a dramatic increase in U.S. methanol use, most of the supply would have to be imported. While biomass-generated methanol might be economical in the long term, there is a significant amount of so-called stranded natural gas in locations around the globe that could be converted to methanol and shipped by tanker at relatively low cost. There would need to be enough natural gas for a growing demand for gas-fired power plants and fuel cells. Methanol from natural gas would have little or no net greenhouse gas benefits in fuel cell vehicles. But, the price of methanol may not remain competitive with gasoline if methanol demand increases. Health and safety concerns would need to be solved and direct methanol fuel cells would need to be affordable.

Liquid Hydrogen

Liquid hydrogen is used today for storing and transporting hydrogen. Liquids have a number of advantages over gases for storage and fueling. They have a high energy density and are easier to transport and handle.

At atmospheric pressures, hydrogen is a liquid at -253°C (-423°F), which is only a few degrees above absolute zero. It can be stored in highly insulated tanks. This cryogenic storage is used by the National Aeronautics and Space Administration (NASA). Liquid hydrogen, along with liquid oxygen has been used as a rocket fuel since World War II.

As a fuel for the space shuttle, almost 100 tons (400,000 gallons) are stored in the shuttle's external tank. To prepare for a shuttle launch requires fifty tanker trucks to drive from New Orleans to the Kennedy Space Center in Florida. There is a great deal of experience in shipping liquid hydrogen. Since 1965, NASA has moved well over 100,000 tons of liquid hydrogen to Kennedy and Cape Canaveral by tanker truck.

Liquid hydrogen can be stored in vessels that are relatively compact and lightweight. General Motors has designed a 90-kg cryogenic tank that holds 4.6-kg (34 gallons) of liquid hydrogen. Liquefying the hydrogen requires special equipment and is very energy-intensive. The refrigeration requires multiple stages of compression and cooling.

About 40% of the energy of the hydrogen is required to liquefy it for storage. The smaller liquefaction plants tend to be more energy-intensive which could be a problem for using local fueling stations.

Another problem with liquefied hydrogen is evaporation. Hydrogen in its liquid form can boil off and escape from the tank. NASA loses almost 100,000 pounds of hydrogen when fueling the shuttle requiring 44% more than the 227,000 pounds needed to fill the main tank. In an automobile, this effect is particularly severe when it remains idle for a few days. The GM tank has a boil-off rate of up to 4% per day. There are techniques for bleeding and using the evaporating hydrogen, but this adds system complexity.

The cost of cryogenic storage equipment which includes the cost of the storage tank and equipment to liquefy hydrogen is high. Liquid hydrogen requires extreme precautions in handling because of the low temperature. Fueling is usually done mechanically with robot arm. Even in large, centralized liquefaction units, the electric power requirement is high. The power needed is 12 to 15 kilowatt-hours (kWh) per kilogram of hydrogen liquefied.

Compressed Hydrogen

Compressed hydrogen has been used in demonstration vehicles for many years, and most prototype hydrogen vehicles use this type of storage. Hydrogen compression is a mature technology and low in cost compared with liquefaction.

The hydrogen is compressed to 3,600 to 10,000 pounds per square inch (psi). Even at these high pressures, hydrogen has a much lower energy per unit volume than gasoline. The higher compression allows more fuel to be contained in a given volume and increases the energy density. It also requires a greater energy input. Compression to 5,000 or 10,000 psi is done in several stages and requires an energy input equal to 10 to 15% of the fuel's energy. Compressing 1-kg of hydrogen into 10,000 psi tanks can take 5-kWh or more of energy.

Compressed hydrogen can be fueled relatively fast, and the tanks can be reused many times. The main technical issues are the weight of the storage tank and the volume needed. Tank weight can be improved with the use of stronger and lightweight materials. Tank volume is improved by increasing the pressure.

Until recently, a 5,000 psi tank was considered to be the maximum allowable. Now 10,000 psi tanks are being built. But even these tanks may need to be several times as large as an equivalent gasoline tank.

The higher pressures also increases costs and complexity requiring special materials, seals and valves. Pressure tanks are usually cylindrical in order to ensure integrity under pressure. This reduces flexibility in vehicle design. Liquid fuel tanks can be shaped according to the needs of the vehicle.

The cost of storage increases with the pressure. The cost of an 8,000 psi storage vessel is several thousand dollars per kilogram of capacity. This can be almost 100 times the cost of a gasoline tank. But, advances in material science and economies of scale should greatly reduce this cost.

Metal Hybrids

Metal hybrids are hydrogen-containing compounds that could be used for hydrogen storage. The hydrogen is chemically bonded to one or more metals and released with a catalyzed reaction or heating. Hybrids can be stored in solid form or in a water-based solution. After a hybrid has released its hydrogen, a by-product remains in the fuel tank to be either replenished or disposed of.

Hybrids may be reversible or irreversible. Reversible hybrids act like sponges, soaking up the hydrogen. They are usually solids. These alloys or intermetallic compounds release hydrogen at specific pressures and temperatures. They may be replenished by adding pure hydrogen.

Irreversible hybrids are compounds that go through reactions with other reagents, including water, and produce a by-product. This by-product may have to be processed at a chemical plant.

Research continues with suboptimal hydrogen release. This is the release of only a part of the stored hydrogen. Refueling can take more than five minutes since some hybrids are slow to absorb hydrogen. Others are slow to release it during use. The chemical process in irreversible hybrids can also be very energy-intensive.

Metal hybrids can hold a large amount of hydrogen in a small volume. A metal hybrid tank may be one third the volume of a 5,000 psi liquid hydrogen tank. Hybrid tanks can take different shapes for vehicle design.

Hybrids are heavy and their storage capacity may be less than 2% by weight. So each 1-kg of hydrogen can require 50-kg or more of tank. A tank with 5-kg of hydrogen could weigh more than 250-kg. This weight reduces fuel efficiency. Many hybrids have a theoretical capacity to store a higher percentage of hydrogen by weight and are a major focus of ongoing research.

Carbon Nanotubes

There has been much work with hydrogen storage in buckyballs or carbon nanotubes. These are microscopic structures fashioned out of carbon. This research indicates a potential storage technique using a combination of chemical and physical containment at very high temperature and pressure.

No storage tank technology has all of the ideal characteristics for commercial applications. It would have to be compact, lightweight, safe, inexpensive, and easily filled.

Compressed gas is well developed in spite of its drawbacks, liquid hydrogen is usable but not widely considered practical and hybrids may be a future technique. A 2003 report by the National Research Council found that compressed hydrogen storage at 5,000 to 10,000 psi would be costly, not only for the storage canisters but also for the compressors and energy needed for compression at refueling stations.

Various technologies may be used, according to the application. Cars, sport utility vehicles, vans, buses, and heavy trucks have different needs and these needs also vary by application, such as urban fleet trucks and long-haul fleets. In some vehicles, volume is more important while in others it may be weight or cost.

Compressed gas is being used in most current demonstration vehicles. But, the path to commercialization of any major new technology is a long one. In 2003 Toyota recalled all six of its hydrogen-powered fuel cell vehicles when a leak was discovered in the fuel tank of one of the cars leased to Japan's Ministry of the Environment. The leak was found when a driver at the ministry heard a strange noise in the car when he was filling up the hydrogen tank. The problem was quickly identified and was fixed a few weeks later.

Biomass Sources

Biomass can include any organic plant or animal matter and biomass energy or bioenergy is a general term for the energy stored in organic wastes. Biomass energy conversion can range from harvesting crops and burning them or distilling their sugars into liquid fuels.

Biomass energy production can replace a variety of traditional energy sources such as fossil fuels in solid or liquid forms. One of the most common sources of biomass energy is wood and wood wastes. Other sources include agricultural residues, animal wastes, municipal solid waste (MSW), microalgae and other aquatic plants. Crops may also be grown for harvesting their energy content. These crops include grains, grasses, and oil-bearing plants. Medium-Btu gas is already being collected at more than 120 landfills in the U.S. Energy farms have the potential of providing an important energy resource.

Biomass technology allows the carbon in the organic matter to be recycled. Unlike the burning of fossil fuels, the combustion of biomass recycles the carbon set by photosynthesis during the growth of the plant. In biomass energy production, the combustion of plant matter releases no more carbon dioxide than is absorbed by its growth and the net contribution to greenhouse gases is zero.

Wood and wood waste includes residues from the forest and the mill. Bark, sawdust and other mill wastes are all suitable fuels. Agricultural residues include corncobs, sugarcane bagasse (the stalks after processing), leaves, and rice hulls. MSW materials include paper products, cloth, yard wastes, construction debris, and packaging materials.

Biomass materials depend on local conditions. In tropical areas, sugarcane is widely grown and bagasse is available as an energy feedstock. Rice growing areas have rice husks available. The Midwestern area of the U.S. can use corn husks and forested areas have timber residues.

Biomass is not a renewable resource unless creation of the source equals or exceeds its use. This is true in energy farms and standard crops, particularly forests.

Biomass Use

Before the use of coal and oil sources became widespread, biomass in the form of firewood was the principal energy source in the U.S. This was also true in most other countries. In the Canada of 1867, biomass was used for 90% of its energy. Only 10% of this nation's energy supply came from other sources such as coal and hydropower.

As coal and then oil became more widespread, the use of biomass dropped, reaching a low point by 1960. Since then, the trend has begun upward with biomass gaining popularity as an energy source. In the forest products industry, wood supplies a large percentage of the energy needed. This is between 65 and 100%, depending on the country.

Biomass accounts for almost 15% of energy production worldwide. In developing countries the amount can be as high as 50%. Nepal, Ethiopia, and Haiti derive most of their energy from biomass. Kenya, Maldives, India, Indonesia, Sri Lanka, and Mauritius derive over half.

In the U.S. about 8% of the energy is provided by biomass. Almost 90% of the biomass energy comes from the combustion of wood and wood residues. The use of biomass increased from an installed capacity of 200 megawatts in 1980 to over 7,700 megawatts in 1990. The search for cleaner fuels and landfill restraints are the main reasons for increased biomass utilization.

The cost of waste disposal has soared and landfill sites are closing faster than new ones are opening up. The Environmental Protection Agency (EPA) estimated that between 1978 and 1988, 70% of the nation's landfills, about 14,000 sites closed.

By the 1990s several states had developed notable biomass energy. Florida's power plants generated more than 700 megawatts of energy from biomass and almost one fourth of Maine's baseload requirements were met with biomass generation. Hawaii generated about one half of its energy from renewable sources and one half of this came from biomass. States with large populations used biomass to help dispose of their waste. Florida, California and New York were large users of MSW for energy.

In Canada, biomass energy equaled the energy produced by nuclear plants and represented about one half of that produced from coal. Biomass made up 12% of the energy in the Atlantic area and almost 25% in British Columbia. In Canada, biomass energy was used for greenhouse heating, health-care facilities, educational institutions, office and apartment buildings, and large industrial plants including automobile manufacturing and food processing. Developed nations that generated higher proportions of their energy from biomass include Ireland with 17% and Sweden with 13%.

Biomass Application

Biomass can generate energy in many different forms. Refuse derived fuels (MSW) can produce steam or electric power. They can also be converted to other fuels using chemical or biological processes producing ethanol or methanol. The wood and pulp industries use their wastes to provide a significant part of their heat, steam, and electricity needs. In Hawaii, the sugar industry produces 150 megawatts of energy from burning bagasse. About half of this energy is sold to the utilities. Mills that process rice may also generate process heat, that can be used for direct heating, steam generation, mechanical power or electrical power.

For every five tons of rice milled, one ton of husks with an energy content equivalent to one ton of wood is left as residue. A rice mill in Louisiana has satisfied all its power needs since 1984 from an on-site rice-husk power plant. The plant sells surplus energy to the local utility.

The first commercial power plant to burn cattle manure to generate electricity was established in the Imperial Valley of southern California in 1987. The plant has a capacity of about seventeen megawatts and supplies electricity to 20,000 homes. The manure is burned to produce steam for the generator.

Biomass feedstocks can also be used to create gaseous and liquid fuels. These can be used on-site, to improve the efficiency of the process or they can be used in other applications. Sugar, starch or lignocellulosic biomass such as wood, energy crops, or MSW can provide alcohols such as methanol, ethanol, and butanol. These fuels can be used as a substitute, or additive, to gasoline. Microalgae and oilseed crops can provide diesel fuel. The use of these alcohol fuels can reduce air pollution. Methane made from anaerobically digested manure was used to light streets in England as early as 1895. Anaerobic digestion is also used to produce fertilizers.

When biomass is transformed into energy by burning, it releases CO_2 that was previously sequestered or held in the atmosphere, for some time, so the net CO_2 emitted is zero. Biomass provides the potential of a sustainable way of providing energy.

Biomass may also be distilled to a liquid fuel, such as ethanol, which is then used to replace gasoline, usually as a blend. Ethanol produced from corn, provides about 25% more energy than that required to grow the corn and distill the ethanol. Ethanol from other sources includes dedicated energy crops such as switchgrass, which may be grown and harvested with less energy consumption.

Biomass Hydrogen

Biomass could be a source of hydrogen. The biomass includes any material that is part of the agricultural growing cycle. Agricultural food, wood and waste products can be used as well as trees and grasses grown as energy crops.

Biomass may be a low cost renewable source of hydrogen in the near future. It could be a major renewable source of hydrogen. Biomass can be gasified and converted into hydrogen and electric power. The process is similar to coal gasification.

Biomass gasification processes are in the demonstration phase and practical commercial systems remain in the future. Biomass can also be gasified together with coal. Royal Dutch/Shell has commercially demonstrated a 25/75 biomass/coal gasifier.

The CO_2 could be extracted from biomass gasification since it is similar to coal gasification. It would mean extracting CO_2 from the air while growing and then injecting that CO_2 into underground reservoirs through the gasification and sequestration process.

Another technique is pyrolysis, which is the use of heat to decompose biomass into its components. This could result in bio-refineries where biomass is converted into many different useful products. The biomass is dried and heated, the co-products are removed and hydrogen is produced using steam reforming.

Biomass may not be feasible for small scale on site hydrogen production. The cost of delivered hydrogen from biomass gasification is estimated to range from $5 to $6/kg, depending on the type of delivery used. Studies by NREL suggest a lower cost, especially for pyrolysis, if the technology is improved. Waste biomass, such as peanut shells or bagasse which is the residue from sugarcane is the most cost-effective source, but the supply is limited.

Even in a country with much arable land such as the U.S., a large part of the agricultural land would be needed for biomass production for it to serve as a major source of fuel. A good fraction of arable land in the United States (and the world) would be needed for biomass-to-hydrogen production sufficient to displace a significant fraction of gasoline which may not be a practical or politically feasible approach. Hydrogen from biomass would also have to be made cost-competitive with gasoline and with other sources of hydrogen. If hydrogen is generated from large biomass plants away from cities, there would be significant infrastructure costs for delivering the hydrogen to consumers. Hydrogen generation from biomass would also have to be a good use for this resource from an environmental and an energy security perspective.

BIOMASS TECHNOLOGY

Plants create energy through photosynthesis using solar radiation and converting carbon dioxide and water into energy crops. Technology allows us to take that energy and transform it through a variety of processes for our uses. The three basic types of bioenergy conversion are direct combustion, thermochemical conversion, and biochemical conversion.

Direct combustion of wood and other plant matter has been a primary energy source in the past. Any type of biomass can be burned to produce heat or steam to turn a generator or perform mechanical work. Direct combustion is used in large power plants that produce up to 400 megawatts. Most direct combustion systems can use any type of biomass as long as the moisture content is less than 60%. Wood and wood residues are commonly used along with a number of other agricultural residues.

Biofuels

Biofuels come from biomass products such as energy crops, forestry and crop residues and even refuse. One characteristic of biofuels is that three fourths or more of their energy is in the volatile matter or vapors, unlike coal, where the fraction is usually less than half. It is important that the furnace or boiler ensure that these vapors burn and are not lost.

For complete combustion, air must reach all the char, which is achieved by burning the fuel in small particles. This finely-divided fuel means finely-divided ash particulates which must be removed from the flue gases.

The air flow should be controlled. Too little oxygen means incomplete combustion and leads to the production of carbon monoxide. Too much air is wasteful since it carries away heat in the flue gases. Modern systems for burning biofuels include large boilers with megawatt outputs of heat.

Direct combustion is one way to extract the energy contained in household refuse, but its moisture content tends to be high at 20% or more and its energy density is low. A cubic meter contains less than 1/30th of the energy of the same volume of coal.

Refuse-derived fuel (RDF) refers to a range of products resulting from the separation of unwanted components, shredding, drying and treating of raw material to improve its combustion properties. Relatively simple processing can involve separation of large items, magnetic extraction of ferrous metals and rough shredding. The most fully processed product is known as densified refuse-derived fuel (d-RDF). It is the result of separating out the combustible part which is then pulverized, compressed and dried to produce solid fuel pellets with about 60% of the energy density of coal.

Anaerobic digestion, like pyrolysis, occurs in the absence of air. But, the decomposition is caused by bacterial action rather than high temperatures. This process takes place in almost any biological material, but it is favored by warm, wet and airless conditions. It occurs naturally in decaying vegetation in ponds, producing the type of marsh gas that can catch fire.

Anaerobic digestion also occurs in the biogas that is generated in sewage or manure as well as the landfill gas produced by refuse. The resulting gas is a mixture consisting mainly of methane and carbon dioxide.

Bacteria breaks down the organic material into sugars and then into acids which are decomposed to produce the gas, leaving an inert residue whose composition depends on the feedstock.

The manure or sewage feedstock for biogas is fed into a digester in the form of a slurry with up to 95% water. Digesters range in size from a small unit of about 200 gallons to ten times this for a typical farm plant and as much as 2000 cubic meters for a large commercial installation. The input may be continuous or batch. Digestion may continue for about 10 days to a few weeks. The bacterial action generates heat but in cold climates additional heat is normally required to maintain a process temperature of about 35°C.

A digester can produce 400 cubic meters of biogas with a methane content of 50% to 75% for each dry ton of input. This is about two thirds of the fuel energy of the original fuelstock. The effluent which remains when digestion is complete also has value as a fertilizer.

A large proportion of municipal solid wastes (MSW), is biological material. Its disposal in deep landfills furnishes suitable conditions for anaerobic digestion. The produced methane was first recognized as a potential hazard and this led to systems for burning it off. In the 1970s some use was made of this product.

The waste matter is miscellaneous in a landfill compared to a digester and the conditions not as warm or wet, so the process is much slower, taking place over years instead of weeks. The product, called landfill gas (LFG), is a mixture consisting mainly of CH_4 and CO_2.

A typical site may produce up to 300 cubic meters of gas per ton of wastes with about 55% by volume of methane. In a developed site, the area is covered with a layer of clay or similar material after it is filled, producing an environment to encourage anaerobic digestion. The gas is collected by pipes buried at depths up to 20 meters in the refuse.

In a large landfill there can be several miles of pipes with as much as 1000 cubic meters an hour of gas being pumped out. The gas from landfill sites can be used for power generation. Some plants use large internal combustion engines, standard marine engines, driving 500-kW generators but gas turbines could give better efficiencies.

Methanol can be produced from biomass by chemical processes. Fermentation is an anaerobic biological process where sugars are converted to alcohol by micro-organisms, usually yeast. The resulting alcohol is ethanol. It can be used in internal combustion engines, either directly in modified engines or as a gasoline extender in gasohol. This is gasoline containing up to 20% ethanol.

One source of ethanol is sugar-cane or the molasses remaining after the juice has been extracted. Other plants such as potatoes, corn and other grains require processing to convert the starch to sugar. This is done by enzymes.

The fuel gas from biomass gasifiers can be used to operate gas turbines for local power generation. A gas-turbine power station is similar to a steam plant except that instead of using heat from the burning fuel to produce steam to drive the turbine, it is driven directly by the hot combustion gases. Increasing the temperature in this way improves the thermodynamic efficiency, but in order not to corrode or foul the turbine blades the gases must be very clean which is why many gas-turbine plants use natural gas.

Biomass Conversion

One biomass conversion unit transforms wood chips into a methane rich gas that can be used in place of natural gas. Another biomass plant in Maine burns peat to produce power. In addition to trees, some smaller plants, like the creosote bush, which grow in poor soil under dry conditions, can be used as sources of biomass, the biological materials that can be used as fuel. These renewable sources of energy can be grown on otherwise unproductive land.

Another type of biomass used as fuel comes from distilleries using corn, sorghum, sugar beets and other organic products. The ethyl alcohol, or ethanol fuel can be mixed in a ratio of 1-to-10 with gasoline to produce gasohol. The mash, or debris, that is left behind contains all the original protein and is used as a livestock feed supplement. A bushel of corn provides two and a half gallons of alcohol plus by-products that almost double the corn's value. Ethanol is a renewable source of energy, but critics question turning food-producing land into energy production.

Thermochemical Conversion

Thermochemical conversion processes use heat in an oxygen controlled environment that produce chemical changes in the biomass. The process can produce electricity, gas, methanol and other products. Gasification, pyrolysis, and liquefaction are thermochemical methods for converting biomass into energy.

Gasification involves partial combustion to turn biomass into a mixture of gases. Gasification processes may be direct or indirect. The direct processes uses air or heat to produce partial combustion in a reactor. Indirect processes transfer the heat to a reactor its walls using heat exchangers or hot sand.

This process produces low or medium Btu gases from wood and wood wastes, agricultural residues and MSW. Processing these synthetic gases with water can produce ammonia, methanol, or hydrogen. Commercial gasification systems exist, but their widespread use has been limited by hauling distances for the feedstock.

Pyrolysis

Pyrolysis is a type of gasification that breaks down the biomass in oxygen deficient environments, at temperatures of up to 400°F. This process is used to produce charcoal.

Since the temperature is lower than other gasification methods, the end products are different. The slow heating produces almost equal proportions of gas, liquid and charcoal, but the output mix can be adjusted by changing the input, the temperature, and the time in the reactor.

The main gases produced are hydrogen and carbon monoxide and dioxide. Smaller amounts of methane, ethane, and other hydrocarbons are also produced. The solids left are carbon and ash. The liquids are similar to crude oil and must be refined before they can be used as fuels.

Liquefaction

In liquefaction systems wood and wood wastes are the most common fuelstocks. They are reacted with steam or hydrogen and carbon monoxide to produce liquids and chemicals. The chemical reactions that take place are similar to gasification but lower temperatures and higher pressure are used. Liquefaction processes can be direct or indirect. The product from liquefaction is pyrolytic oil which has a high oxygen content. It can be converted to diesel fuel, gasoline or methanol.

Biochemical Conversion

Biochemical conversion, or bioconversion, is a chemical reaction caused by treating moist biomass with microorganisms such as enzymes or fungi. The end products may be liquid or gaseous fuels. Anaerobic digestion and fermentation are the two processes used for biochemically converting biomass to energy.

Anaerobic digestion involves limiting the air to moist biomass such as sewage sludge, MSW, animal waste, kelp, algae, or agricultural waste. The feed stock is placed in a reaction vessel with bacteria. As the bacteria break down the biomass, they create a gas that is 50 to 60% methane. Small scale digesters are used on Asian and European farms. Sewage treatment plants use this process to generate methane and digesters are used to compost municipal organic waste.

Anaerobic systems range from large systems that can handle 400,000 cubic feet of material and produce 1.5 million cubic feet of biogas per day to small systems that handle 400 cubic feet of material and produce 6,000 cubic feet of biogas a day.

Fermentation

Fermenting grains with yeast produces a grain alcohol. The process also works with other biomass feedstocks. In fermentation, the yeast decomposes carbohydrates which are starches in grains, or sugar from sugarcane juice into ethyl alcohol (ethanol) and carbon dioxide. The process breaks down complex substances into simpler ones.

Ethanol is a healthy industry in some parts of the United States and the rest of the world. It is an alternative as an automobile fuel. Brazil has a large ethanol industry, producing about three billion gallons each year from sugarcane.

BIOMASS POTENTIAL

Developing biomass energy can provide economic, political, social and environmental advantages. The energy potential of biomass has been estimated at almost 42 quadrillion Btus which is about 1/2 of the total energy consumption in the United States. Biomass provides the U.S. with about the same amount of energy as the nuclear industry. Biomass can provide substitutes for fossil fuels as well as electricity and heat. Its resource base is varied. Arid land, wetlands, forest, and agricultural lands can provide a variety of plants and organic matter for biomass feedstock.

Converting waste products to energy lowers disposal costs and provides cost savings in purchasing energy supplies. Profitability can be improved by using waste to create energy. The sugar industry converts bagasse to energy and sells excess power. Biomass facilities often require less construction time, capital, and financing than many conventional plants.

In the Northeast alone, biomass accounts for over $1 billion in the economy and almost 100,000 jobs. Biomass production offers crop alternatives and the potential for increased income to farmers. Fields that are not used in winter can produce biomass, and varying crops in the same fields can help protect soil quality.

Biomass energy offers an increased supply with a positive environmental impact. If grown on a sustainable basis, it causes no net increase in carbon dioxide and the use of alcohol fuels reduces carbon monoxide emissions. Biomass is renewable as long as it is grown on a sustainable basis.

Although the feedstocks are widespread, they must be used locally since their bulk makes it costly to transport the feedstocks. In California, it has been uneconomical to transport wood residues more than 100 miles. The bulkiness of biomass resources can also cause storage problems.

Many available biomass feedstocks have a high moisture content, which lowers their heat value. Preprocessing can help, but adds to the cost. There are also some biomass conversion technologies that are only marginally beneficial and this keeps them from being cost-competitive. Table 7-2 shows the major sources of biomass utilization in the United States.

Table 7-2. Utilization of Biomass Resources in the U.S.

Wood and wood waste (industrial/commercial)	50%
Wood (residential space heating)	35%
MSW, agricultural wastes, landfill gas and biogas, alcohol fuels	15%

A major increase in biomass energy production has the potential to cause serious environmental problems. Land use issues and concerns about pollution are significant concerns. Areas with fragile ecosystems and rare species would need to be preserved. Agricultural lands would also compete with food production. The loss of soil fertility from overuse is another issue. Biomass energy production would need to be varied and sustainable while preserving local ecosystems.

Pollution problems could result from the increased use of fertilizers and bioengineered organisms on energy farms. The introduction of hazardous chemicals from MSW into the agricultural system could result in increased air and water pollution.

Cost Issues

The usual goal for installing other energy systems in industries or institutions is to achieve a net savings in energy costs. These savings are realized when the energy costs of the sources being replaced are more than the total operating and installation costs of the biomass system.

Greenhouses, lumber mills, canneries, farmers, and manufacturers can reduce energy and disposal costs by using their waste as feedstock for energy systems. In Ireland, greenhouses for early tomatoes are heated with biomass from willow wood. The willow wood fuel costs one third as much as the oil it replaced.

One study of biomass used in Honduras investigated an energy-efficient power plant that used all the wastes of a large lumber mill. It sold power to the grid for at least $0.05 per kilowatt hour and could produce an internal rate of return on equity investment of 75%, and paid back the initial investment in just over three years.

The United States Agency for International Development studied the use of sugarcane residues for power in Thailand, Jamaica, the Philippines and Costa Rica. The study found that cane power can have lower unit costs than most of the other power generation options available to these countries studied. In Thailand the study found that a new cane power plant could supply power at about $0.030 per kilowatt hour. This was well below the cost of power generated in that country with imported coal at $0.044 per kilowatt hour or domestic natural gas at $0.040 per kilowatt hour.

Another study by the California Energy Commission found that wood-fired boilers can be installed for about $1,340 per kilowatt, which is 20% less than a coal unit. Biomass conversion plants are often smaller than fossil fuel units and can be built more quickly, less expensively and with less capital investment.

Biomass wastes offer a cost-effective energy alternative, but energy plants grown specifically for energy production may not be competitive with fossil fuels. New biotechnologies are needed to improve the energy production in crops, along with new combustion technologies and more efficient gas turbines.

The Future for Biomass

Government R&D funding for biomass increased slightly by 1990, but had decreased more than 75% from 1980 to 1989. The U.S. Department of Energy projected in 1989 that biomass could potentially become the world's largest single energy source if intervention to protect the climate takes place.

France has been experimenting with short-rotation forestry on more than 400 hectares of land. Northern Ireland is conducting similar experiments. India has expanded its network of biogas digesters, which supplies compost to farmers and power to local communities. Finland provides almost 20% of its energy needs from biomass is working to increase this to 35% through using forest and peat feedstocks. Exotic fuels such as those derived from algae may also be used but these still need more development. Other areas that need development include micro organisms for anaerobic digesters, genetic engineering for superior microbes, yeasts, and fungi, catalytic processing of lignins to liquid fuels and advanced fermentation techniques.

Increased use of MSW as a fuel is also expected with the United States currently producing over 200 million tons of garbage each year. MSW offers a large, growing resource, even after recyclables are removed.

While it may be theoretically possible to replace the use of fossil fuels around the world with biomass energy, it is more likely, that biomass will take on a more important role as one of our energy sources.

WIND POWER

During the development of electrical generating equipment in the late 1800s, both Europe and America began to experiment with wind power for electrical generation. Among the first to develop wind-powered electrical generators was the Danish professor, Poul La Cour, who worked on wind systems from 1891 to 1908. He also saw the use of hydrogen as a fuel and the use of wind-powered electrical generators to electrolyze hydrogen and oxygen from water. Another early investigator who promoted wind-powered hydrogen production systems was J.B.S. Haldane a British biochemist at Cambridge, England. In 1923, he predicted that England's energy problems could be solved with a large number of wind generators supplying high voltage power for hydrogen production.

During World War II, Vannovar Bush was the Director of the U.S. Wartime Office of Scientific Research and Development. He was concerned about American fuel reserves and thought that wind generators could be a solution. Percy Thomas was a wind power advocate on the Federal Power Commission, who convinced the Department of the Interior to construct a large prototype wind generator. In 1951, the House Committee on Interior and Insular Affairs killed this plan. Wind-generated electricity could not compete with coal that was selling for $2.50 per ton or diesel fuel at $0.10 per gallon. The promise of even less expensive electricity that was too cheap to meter from nuclear power plants resulted in the loss of almost all Federal programs to develop wind-powered energy systems.

Today's wind machines are known as wind turbines and can have rotors that slash through the air at heights of up to 100 meters. More and more of these giant machines are being installed around the world. Wind power only provides 0.15% of the world's total electricity, but it has become the fastest growing form of energy production.

Wind Turbines

The basic principles of wind energy have been used for centuries. Windmills existed in the 7th century in Persia. An older image closely associated with wind power is Don Quixote and the wooden towers with cloth-covered sails turning in the wind. Today's wind turbines use a giant propeller on a tall metal pole. As it rotates, the propeller drives a generator to supply nearby users or send power to the grid.

The trend is for wind farms to move offshore, where their appearance and the sound of whirring propellers will not bother local communities. The strong and steady sea winds keep the turbines turning most of the time.

For the past few decades, manufacturers have been streamlining components and installing onboard computers to tilt the propeller blades for maximum efficiency for the wind conditions. In the 1980s, the average turbine was 20 meters high with a 26-kilowatt (kW) generator and a rotor diameter of 10.5 meters. A typical turbine today can be 55 meters high, with a rotor diameter of 50 to meters and a capacity of 1,650-kW. The power it produces can supply almost 400 homes.

Since 1992, more commercial wind farms have been installed than ever before with 40,000 turbines in 40 countries. Wind energy capacity is growing at almost 30% annually. By 1998, it reached 10,000 megawatts (MW), which can supply a country the size of Denmark and the wind power industry had sales of $2 billion with 35,000 jobs worldwide. The prime movers were an increasing environmental awareness and commitments to reduce greenhouse gas emissions made under the Kyoto Protocol of 1997.

The European Union supplies tax breaks and investment plans for renewable sources such as wind power. There are plans to install 40,000 megawatts by 2010. Denmark receives 10% of its power from wind energy with an installed capacity of 1,700-MW. Germany is not far behind and is the wind sector's fastest growing market.

Wind power in the U.S. has not received this level of support. Every two years, a fight in congress erupts on the renewal of an important wind power tax credit. Similar battles occur in state legislatures that have wind power tax credits. According to U.S. energy officials, wind power should provide 5% of the nation's electric power by 2020, compared to the current 0.1%.

Public Combat

Wind power is being slowed by public opposition. In 2002, a citizen's group in Prince Edward County, Ontario, vetoed a small wind farm project on the coast of Lake Ontario near Hillier. They thought that the 22 proposed wind turbines would be noisy, kill birds and harm the neighborhood by being too visible. These are common complaints about wind farms, but at a distance of about 200 meters, the sound of a wind farm is faint. Closer the noise is similar to the sound of an airplane's engine from inside the cabin. Even under the spinning blades it is possible to converse in a normal voice. One Dutch study showed that a small wind farm is less harmful to birds than 1-kilometer of road or powerlines.

The U.K. has the best wind resources in Europe, but attempts to set up wind farms were stopped when local authorities failed to issue permits for turbine construction. The national government had no guidelines and policies allowing local authorities to cooperate.

In India and China, wind power can provide broad areas of the rural population that are without electricity. Wind investment plans are being offered to these countries by Denmark, Germany and the Netherlands. India has almost 850-MW of installed capacity and is first among the developing countries and fourth in the world after Germany in wind power. About 600 wind turbines are producing 260-MW in China. New Zealand has its Tararua Wind Farm which is the largest in the southern hemisphere with a capacity of 12-MW.

In North Africa, Morocco recently installed 50- MW and Egypt 30-MW. Wind power could provide at least 20% of every continent's energy needs. There is enough wind to provide twice the expected global power demand for 2020.

If 10% of energy needs were met by wind power, there would be about 10 billion tons less of worldwide carbon emissions out of a world total of 60 to 70 billion tons. To achieve this, 120 times more wind capacity is needed.

Wind Power Costs

Initial investments are high, but operation and maintenance costs for wind power are low. Bigger and better turbines have resulted in wind power prices dropping by about 20% over the past several years. In Denmark, wind power costs were almost 17 cents per kilowatt hour (kWh) in the early 1980s. This includes equipment, labor, interest on loans, operation and maintenance. It dropped to about 6.1 cents by 1995 and was 4.5 cents by 2001.

Power from a new coal-fired power plant will cost 5 to 6.4 cents per kWh and 4 to 5.7 cents per kWh for a gas-fired plant, and 4.6 to 6.5 cents per kWh for a nuclear plant, according to UNIPEDE, the European Utility Association.

The cost of wind-powered electricity should continue to fall in the future. The steep start-up costs of installing wind turbines are the downside of wind power.

References

Cothran, Helen, Book Editor, *Global Resources: Opposing Viewpoints*, Greenhaven Press,: San Diego, CA, 2003.

Romm, Joseph J., *The Hype About Hydrogen*, Island Press: Washington, Covelo, London, 2004.

CHAPTER 8

ALTERNATIVE FUEL PATHS AND SOLAR HYDROGEN

The road to hydrogen vehicles and a hydrogen fueling delivery system may take many paths. Today, it may seem unlikely that market forces alone will result in the installation of thousands of hydrogen fueling stations spread uniformly across the country. But, this is exactly what happened with our present oil economy. Gasoline was originally available in small amounts often from hand pumps. As demand for gasoline for automobiles grew, so did fuel outlets.

An examination of efforts by the federal government to promote alternative fuel vehicles in the 1990s illustrates the lack of interest in alternative fuels when gasoline is widely available. In 1992, the United States passed the Energy Policy Act. One goal was to reduce the amount of petroleum used for transportation by promoting the use of alternative fuels in cars and light trucks. These fuels included natural gas, methanol, ethanol, propane, electricity, and biodiesel. Alternative fuel vehicles (AFVs) can operate on these fuels and many are dual fueled also running on gasoline.

Another goal was to have alternative fuels replace at least 10% of petroleum fuels in 2000 and at least 30% in 2010. Part of the new vehicles bought for state and federal government fleets, as well as alternative fuel providers, must be AFVs. The Department of Energy (DOE) was to encourage AFVs in several ways, including partnerships with city governments and others. This work went to the Office of Energy Efficiency and Renewable Energy. The initial efforts were reported in a 2000 report by the General Accounting Office (GAO).

By 1999, some 1 million AFVs were in use which is less than 0.5% of all vehicles. In 1998, alternative fuels used by AFVs replaced almost 335 million gallons of gasoline, about 0.3% of the year's total consumption. Almost 4 billion gallons of ethanol and methanol replaced gasoline that year in blended gasoline that was sold for standard gasoline engines.

The DOE has been developing clean energy technologies and promoting the use of more efficient lighting, motors, heating and cooling. As a result of these efforts and efforts by others, there have been savings by business and consumers of more than $30 billion in energy costs. Getting people to use alternative fuel vehicles has proven to be more difficult.

The GAO has stated that goals in the act for fuel replacement are not being met because alternative fuel vehicles have serious economic disadvantages compared to conventional gasoline engines. These include the relatively low price of gasoline, the lack of refueling stations for alternative fuels and the additional costs of these vehicles.

HYDROGEN AND INTERNAL COMBUSTION

Hydrogen powered internal combustion engines could promote infrastructure for fuel cell cars. An internal combustion engine (ICE) can burn hydrogen with a few inexpensive modifications. Automakers, including Ford and BMW, have been planning to introduce hydrogen ICE cars. They have the advantage over gasoline engines of very low emissions of urban air pollutants. However, there is the relatively high cost of today's hydrogen. Hydrogen engines are also about 25% more efficient than gasoline units. They are likely to have a smaller driving range due to the problem of storing large volumes of hydrogen on board.

Due to the high price of hydrogen, annual vehicle costs for mid-sized hydrogen vehicles could be one third higher than for current gasoline vehicles. This is slightly lower than the estimated annual costs for fuel cell vehicles, according to a report by the Arthur D. Little firm.

Because of the energy used in generating hydrogen from natural gas or electricity and the energy required to compress hydrogen for storage, the total energy use of a hydrogen internal combustion engine can be higher than a gasoline engine. One study of ten different alternative fuel vehicles found that burning hydrogen from natural gas had the lowest overall efficiency on a total energy consumed basis.

A report by the U.S. General Accounting Office (GAO) found that officials from federal agencies and state governments pointed to the lack of a refueling infrastructure more than any other reason to avoid alternative fuels.

Fleet Use

Fleet use is another strategy for alternative fuel commercialization. It was the main strategy that the DOE used in the 1990s to meet the goals of the Energy Policy Act of 1992. Vehicle fleets are typically driven twice as many miles compared to private vehicles.

Fleet vehicles make up about one fourth of all U.S. light-duty vehicle sales. Many fleet vehicles have fixed daily routes and are regularly fueled at one location, so less infrastructure is needed to support fleet-based vehicles.

One survey of almost 3,700 California fleets, found two main reasons why central fueling may actually be a problem for alternate fuels. Light-duty fleets often reduce fuel costs by purchasing petroleum in bulk. But, hydrogen is currently more expensive than gasoline on an equivalent energy basis.

High travel demands do not match well with fuels that have shorter ranges and limited refueling stations. Gasoline or diesel vehicles provide a longer driving range and can also refuel at commercial gas stations.

Almost 80% of public fleets use central refueling, but only about one third of business fleets do and most of those also use commercial fueling stations. Most fleets that centrally refuel use outside sources for at least 15% of their refueling.

The Environmental Protection Agency (EPA) has had major concerns over fuel leakage and underground water contamination in the last few decades. This has resulted in a significant reduction in the number of underground fuel storage tanks.

In the current era of high federal budget deficits and tight state budgets, it remains to be seen if taxpayers and legislators are willing to spend scarce dollars to subsidize hydrogen vehicles and hydrogen fueling. We have already seen a number of states dip into funds that were dedicated to promote clean energy, using them for other purposes and creating a bad precedent for the future.

Emission and Efficiency

The internal combustion engine, running on gasoline, has been powering transportation for almost a century. Advances in engines and fuels, such as reformulated gasoline, have reduced the pollution of these engines.

Competitors such as electric cars and natural gas vehicles have not been able to penetrate their dominance. The competition for fuel cell vehicles includes hybrid vehicles and diesels, which are seeing many advances today.

Hybrid gasoline electric-powered cars can be twice as efficient as internal combustion vehicles. An onboard energy storage device, which is usually a battery and sometimes a special capacitor (called a super capacitor), increases the efficiency greatly. Regenerative braking is also used to capture energy that is normally lost when the car is braking. The engine is turned off when the car is idling or decelerating. Gasoline engines have lower efficiencies at lower rpm so the gas engine operates only at higher rpms and is more efficient more of the time. In city driving, non polluting electric power is used.

The first-generation Toyota Prius had a city mileage of 52 miles per gallon (mpg) and a highway mileage of 45 mpg. The second-generation Prius, appeared in 2003 with improved mileage numbers. Toyota will be introducing other hybrid models in the future and most auto manufacturers plan to produce hybrid vehicles.

Fuel cells typically have higher efficiencies at lower power, so a hybrid fuel cell vehicle with battery will not improve its efficiency as it does for a gasoline engine.

The high efficiency that hybrids have in urban settings could be particularly tough competition for fuel cell vehicles because, at least initially, fuel cell vehicles are likely to be used mainly for urban driving. Early models probably will not have the driving range of regular vehicles and will be used by fleets, which operate mainly in cities. The limited number of fueling stations early on will restrict long-distance travel.

Diesel

Another competitor for fuel cell vehicles could be the diesel engine. Diesel engines are used in large trucks and construction equipment for their high efficiency and durability. Modern diesel engines are much different from the engines of the 1970s and 1980s. Advances have included electronic controls, high-pressure fuel injection, variable injection timing, improved combustion chamber design, and turbo-charging.

Diesels make up less than 1% of car and light truck sales in the U.S. But, they are more popular in Europe with its high gasoline taxes. Their fuel taxes help to promote diesels and the emissions standards are less strict. Diesels are in almost 40% of the cars in Europe. By 2001 they were in most of the new cars sold in many European countries.

They are 30 to 40% more fuel efficient than gasoline vehicles. The production and delivery of diesel fuel releases 30% less carbon dioxide than producing and delivering gasoline with the same energy content.

Diesels emit higher levels of particulates and oxides of nitrogen. But, they are steadily reducing these emissions. A large amount of R&D is currently going into at diesels and it is expected that they will be able to meet the same standards as gasoline engines in the near future.

Cost Issues

A 2002 report for the DOE estimated that even with technology improvements, future fuel cell vehicles could cost 40 to 50% more than conventional vehicles. Hydrogen may also be much more expensive than gasoline. Hydrogen provided at fueling stations could probably cost as much as $4 or more per kilogram (kg). The equivalent-energy price of gasoline is about 75% of this. A kilogram of hydrogen has almost the same energy as a gallon of gasoline. Ultimately, if hydrogen were to be the main transportation fuel, it would itself have to be taxed unless we find a new source for funding road projects. So, even with a more efficient engine, the annual fuel costs are likely to be higher for fuel cells at least in the beginning.

While hybrid and clean diesel vehicles may cost more than current internal combustion engine vehicles, their greater fuel efficiency means that, unlike hydrogen fuel cell vehicles, they may make up that extra upfront cost over the lifetime of the vehicle. This means that hybrids and diesels may have roughly the same annual operating costs as current internal combustion engine vehicles.

This also means that hybrids and diesels could reduce transportation CO_2 emissions at a lower cost per ton. The typical new car today generates about four to five metric tons of CO_2 per year. One reason for replacing gasoline engines is lower that number. A fuel cell vehicle in 2020 might reduce CO_2 emission at a cost of more than $200 per metric ton, regardless of how the hydrogen was produced. An advanced gasoline engine could probably reduce CO_2 at lower cost.

Tanker Trucks

Tanker trucks with liquefied hydrogen are typically used to deliver hydrogen today. This is the method NASA uses. It is popular for delivery in Europe as well as North America and works to supply distributed users with moderate hydrogen needs. It is currently less expensive than small on-site hydrogen generation and provides high purity hydrogen for industrial processes. Liquefaction has a high energy cost, requiring about 40% of the usable energy in hydrogen. A few automakers are using onboard storage with liquid hydrogen in prototype test vehicles.

Liquid tanker trucks could be the least expensive delivery option in the near future. After delivery, the fueling station still has to use an energy-intensive pressurization system, which can consume another 10 to 15% of the usable energy in the hydrogen. This could mean that storage and transport alone might require as much as 50% of the energy in the hydrogen delivered. For liquefaction to be viable, a less energy-intensive process is needed.

Hydrogen Pipelines

Pipelines can be used for delivering hydrogen. Today, several thousand miles of hydrogen pipelines exist around the world, with several hundred miles in the U.S. These lines are short and located in industrial areas for large users. The longest pipeline in the world is almost 250 miles long and goes from Antwerp to Normandy. It operates at 100 atmosphere of pressure which is approximately 1,500 psi.

Air Products plans on constructing a new hydrogen production plant in Port Arthur, Texas to supply 110 million standard cubic feet per day of hydrogen to Premcor Refining and others on Air Product's Gulf Coast hydrogen pipeline system. Pipelines may be the least expensive way to deliver large quantities of hydrogen. Pipelines are the main choice for moving refined petroleum products across the country. They are less that 10% the cost of rail, road or water tankers.

The U.S. has almost 200,000 miles of interstate pipelines for petroleum products. There is another 200,000 miles of interstate natural gas pipelines.

Hydrogen pipelines are expensive because they must have very effective seals. Hydrogen is also reactive and can cause metals, including steel, to become brittle over time. Hydrogen pipelines of 9 to 14 inch diameter can cost $1 million per mile or more. Smaller pipelines for local distribution cost about 50% of this.

Siting major new oil and gas pipelines is often political and environmentally litigious. Political pressures may favor one location over another. Whether global warming concerns will be enough to override other considerations is still unknown. Pipelines are more likely to be used for hydrogen transport once there is real demand.

Trailers carrying compressed hydrogen canisters provide a flexible way of delivery suited for the early years of hydrogen use.

This is a relatively expensive delivery method since hydrogen has a low energy density and even with high-pressure storage, not that much hydrogen is actually being delivered. Current tube or canister trailers hold less than 300-kg of hydrogen which is enough to fill sixty fuel cell cars. It is estimated that with improved high-pressure canisters, a trailer could hold about 400-kg of hydrogen or enough for about 80 fuel cell cars. A tanker truck for gasoline delivers about 26 metric tons of fuel, or 10,000 gallons which is enough to fill 800 cars.

About one in 100 trucks is a gasoline or diesel tanker, replacing liquid fuels with hydrogen transported by tube truck means that about 10% of the trucks on the road would be transporting hydrogen. Technology may provide better options in the future. There is significant R&D going into each of the storage and transportation technologies.

Iceland may become the world's first hydrogen economy. This island nation in the North Atlantic is the size of Kentucky. It is near the Arctic Circle, but is warmed by the Gulf Stream. Iceland has more of its share of active volcanoes, hot springs, and geysers than any other area of its size in the world. Its 280,000 people are less than 10% of Kentucky's population. Iceland is suited to a hydrogen economy because it has excess renewable energy.

Iceland uses its renewable energy for power generation and heating. So, these sectors are nearly carbon-free, an unusual example. Carbon dioxide emissions are produced by the transportation, fishing, and industrial sectors, each of these contributes about one million tons of carbon dioxide (CO_2) per year.

Using some of its renewable energy would allow Iceland to produce hydrogen and replace all the oil used for the country's transportation and fishing industry. There would still be emissions from industrial processes such as aluminum and ferrosilicon production. But, this plan would cut the country's fossil fuel use dramatically.

Iceland has little fossil fuel resources but has plenty of inexpensive, clean hydropower as well as geothermal energy. It is above an active geothermal area where the continental plates meet. The name of the island's capital, Reykjavik, is based on the hot steam fumes which can be seen along the horizon.

Energy is tapped from the hot water or steam in the ground to run turbine generators. Lower temperature water is used to heat buildings or provide process heat for industries. Geothermal energy is used in 90% of the buildings for hot water or steam. Almost 9 million megawatt-hours (MWh) of thermal energy is used each year for heating and industrial uses. Geothermal heating for buildings provides low cost heat with no need for heat pumps or furnaces.

Iceland also has about 170 megawatts (MW) of geothermal electric power generation which provide more than 1.3 million MWh per year. The total geothermal resource base is about 30,000-MW which is 20 times the existing use.

Geothermal energy can be extracted faster than it is recharged giving many fields limited life times. Energy production can drop as the reservoir loses water or heat. At the Geysers in northern California, geothermal power production dropped quickly in the 1980s and slower declines have appeared at other sites.

Differences in geology indicate that Iceland's geothermal reservoirs may be large enough to sustain the present rate of geothermal energy usage. Studies show that this rate could even be increased.

The amount of energy stored in Iceland's bedrock is estimated at over 25 trillion MWh within three kilometers of the surface. About 1 trillion is considered usable and may be continually recharged from below. This heat could be extracted from reservoirs and converted to energy for a few hundred years at current usage levels. Reservoir heat is replenished by volcano activity which pushes the magma to the surface.

Iceland also has large amounts of mountains, glaciers and precipitation. Its hydroelectric plants have a capacity of approximately 1,000-MW and supply almost 7 million MWh per year of electric power.

Because of environmental issues, Iceland may find it hard to further develop its hydropower reserves. But, the current capacity at hydroelectric plants would allow significant hydrogen production.

The hydrogen could be produced during non-peak hours and stored until it is needed. This would allow Iceland to replace almost one fourth of the fossil fuels consumed by vehicles and vessels using its present generating capacity.

Iceland could also develop wind power with coastal or offshore facilities. A study indicated that 240 wind power plants could produce the electricity needed to replace fossil fuel from vehicles and fisheries.

Other studies suggest that only 17% of Iceland's renewable energy has been developed. This renewable electricity has been estimated at up to 50 million MWh per year for hydropower and geothermal. This represents six times the current renewable energy capacity.

Iceland's Energy Progress

In 1978, Professor Bragi Arnason proposed that Iceland develop hydrogen. Support grew in the 1990s, because of advances in fuel cell technology and concerns about climate changes and a dependence on oil. By 1999, Shell, DaimlerChrysler, Norsk Hydro, an Icelandic holding company Vistorka hf (EcoEnergy) and others created the Icelandic Hydrogen and Fuel Cell Company, now called Icelandic New Energy Ltd. This group with the backing of the government and the European Union started the path to hydrogen in Iceland.

Almost 65% of the population lives near the capital of Reykjavik. This allows a hydrogen infrastructure to be established with a few fueling stations in Reykjavik and nearby connecting roads. In 2003, Iceland opened the first public hydrogen filling station in the world, even though there were no privately owned hydrogen vehicles in the country.

A 2001 survey found that almost 95% of the population supported replacing traditional fossil fuels with hydrogen. Iceland shifted it power generation from fossil fuel to hydroelectric power very early and went to geothermal heating after World War II.

To fuel its vehicles and fishing fleet, Iceland imports about 6 million barrels per year of petroleum. Gasoline and diesel fuel cost over $1 per liter or about $4.00 per gallon. There are no sources of oil or other fuels other than some landfill methane on the island.

In spite of its carbon-free electric power and the widespread use of geothermal heating, Iceland has high CO_2 emissions per capita. Typical developed countries emitted about 12 metric tons per capita in 1990, whereas Iceland emitted about 8.5 metric tons per capita. All forms of energy including renewables can affect the environment. Geothermal power can produce some emission of CO_2, about 100 grams (g) per kilowatt-watt (kWh), which is roughly 30% of the emissions of an efficient combined cycle natural gas plant.

Increasing hydropower production may require additional dams. But, Iceland has unused capacity at existing hydroelectric power plants, which could be used to produce hydrogen in off-peak times.

Hydrogen Transition

Icelandic New Energy proposed a six-phase plan for the hydrogen transition. Phase 1 was under way with the opening of a hydrogen fueling station in 2003. Three fuel cell buses which are 4% of the city's bus fleet have been in use in Reykjavik. This is known as ECTOS for Ecological City Transport System. Phase 2 will replace the Reykjavik city bus fleet with proton exchange membrane (PEM) fuel cell buses. Phase 3 will begin the use of PEM fuel cell cars, while phase 4 will demonstrate PEM fuel cell boats. Phase 5 will replace the entire fishing fleet with fuel cell powered boats and in the next phase Iceland will sell hydrogen to Europe and elsewhere. The last phase is expected to be completed by 2030-2040.

Iceland may start with methanol powered PEM vehicles and vessels. The University of Iceland is researching the production of methanol (CH_3OH) from hydrogen combined with carbon monoxide (CO) or CO_2 from the exhaust of aluminum and ferrosilicon smelters. This would capture hundreds of thousands of tons of CO and CO_2 released from these smelters. If this is combined with hydrogen generated from electrolysis using renewable power, Iceland could cut its greenhouse gas emissions in half.

RENEWABLES

Renewable energy has many attributes similar to those of fuel cells, including zero emission of urban air pollution, but some believe renewable sales have been slowed in the United States because of their high cost. Actually, renewable technologies have succeeded in meeting most projections with respect to cost. As costs have dropped, successive generations of projections of cost have either agreed with previous projections or have been less.

Renewables should become important parts of the power generation mix in the U.S. They exemplify an important long-term success for government R&D.

Government R&D funding for renewables has been exceedingly successful, bringing down the cost of many renewables by a factor of ten in two decades, even though the R&D budget for renewables was cut by 50% in the 1980s and did not rebound to similar funding levels until the mid 1990s.

Wind power has become a major part of power generation in Europe, with 20 to 40% of power loads in parts of Germany, Denmark, and Spain.

Photovoltaics has made much progress, but has had to compete with declining price scales in conventional generation. Traditional electricity generation costs dropped in the 1980s and 1990s rather than increasing, as had been projected in the 1970s. It did this while reducing emissions of urban air pollutants.

Utilities were also allowed to place many barriers in the path of new projects while new technologies typically received little appreciation for the contributions they make in meeting power demand, reducing transmission losses or improving the environment. However, the competition from renewables does push the utilities to improve their performance.

A major area of R&D conducted by DOE's office of Energy Efficiency and Renewable Energy involves energy-efficient technologies that reduce energy bills. More efficient devices include refrigerators, light bulbs, solid-state ballasts for fluorescent lights and improved windows. Many of these products have achieved significant market success. The National Academy of Sciences states that they saved the U.S. $30 billion worth of energy.

The products that were most successful had a good payback combined with similar or superior performance. Solid state ballasts can reduce energy use in half or more while providing a high quality light without the flicker of earlier fluorescent. They can provide a payback of less than two years.

WAVE ENERGY

Wave energy is one of the most promising renewable sources in maritime countries. A wave travels forward in an up-and-down motion and its height is an indication of its power. Ocean waves could soon be providing large amounts of power for maritime countries. In spite of the potential of the sea to destroy wave-energy stations, several nations have made great progress in designing more rugged small-scale wave power stations. A wave power station must be able to absorb the power of the largest waves without being damaged. Two wave power stations, one in Scotland and one in Norway, have already been victims of high waves. The energy potential is great. It has been estimated as being as much as 4,000 gigawatts (GW).

Energy from the waves was studied at the time of the French Revolution when a patent was filed in Paris by a father and son team named Girard. Their patent noted how the enormous mass of a ship of the line, which no other known force was able to lift, responded to the slightest wave motions. There has not much progress in turning this motion into useful energy until the last quarter century.

One recent advance is the oscillating water column (OWC). This is a column that sits on the seabed and admits the waves through an opening near to its base. As the waves rise and fall, the height of the water inside also rises and falls. As the water level rises, air is forced up into a turbine which drives a generator. As the wave drops, the column takes air in from the atmosphere and the turbo-generator is also activated.

Professor Alan Wells of Queen's University, Belfast has developed a turbine which spins in the same direction regardless of direction of the air flow.

Norway built a wave energy station on the coast near Bergen in 1985. It combines an OWC with a Norwegian device called a Tapchan (TAPered CHANnel). The waves move up a concrete slope to 3 meters above sea level, where they fill a reservoir. As the water flows back to the ocean, it drives a turbine generator.

Small-Scale Wave Energy

Small-scale wave power ranging from 100 kilowatts (kW) to 2 megawatts (MW) are now in more than a dozen countries. Scotland had a trial 75-kW OWC on the island of Islay for 11 years. This has been replaced by a 500-kW unit. The same group plans a 2-MW seagoing device called the Osprey. Portugal has been working on an OWC on the island of Pico in the Azores.

An American company is working on a 10-MW system based on buoys 3 kilometers off the south coast of Australia. China, Sweden and Japan are also working on wave energy.

Wave energy is a capital-intensive technology, with most of the costs for construction. But, for the first time for 30 years, a major breakthrough is in sight. Wave electricity should be on the grid in many countries before long.

Another way to harness the sea's energy is to use the difference between sea levels. In Egypt, water running through an underground canal linking the Mediterranean to the El-Qattar depression could be used to generate electricity. In Israel, the same principle could be used in a canal between the Mediterranean and the Dead Sea which would descend 400 meters.

HYDROGEN FROM PHOTOCATALYTIC WATER SPLITTING

The cleavage of water to form hydrogen and oxygen could be a source of hydrogen. The water splitting reaction is endothermic and the energy required to achieve a significant hydrogen production rate is high. Ideally, the energy source should be in abundant supply and non-polluting. Solar energy meets these requirements, and the use of solar energy to drive the cleavage of water to produce hydrogen would be an attractive way to convert solar energy to chemical energy.

The photocatalytic process uses semiconducting catalysts or electrodes in a photoreactor to convert optical energy into chemical energy. A semiconductor surface is used to absorb solar energy and act as an electrode for splitting water. The technology is still at an early stage of development and the most stable photoelectrode is TiO_2. But, this material has a conversion efficiency of less than 1%. New materials, which require no external electricity, may be found. In order to reduce corrosion, very thin layers of protective material on the semiconducting surface may be used. Other areas of research include multiple layers of organic dyes and thin film semiconductors.

Thermochemical Water Splitting

Hydrogen may also be produced by a water-splitting thermochemical cycle based on metal oxides. The simplest thermochemical process to split water involves heating it to a high temperature and separating the hydrogen from the equilibrium mixture. The decomposition of water does not proceed well until the temperature is about 4700K. Problems with materials and separations at such high temperatures makes direct decomposition difficult.

Hydrogen production by a 2-step water splitting thermochemical cycle can be based on metal oxides redox pairs. A two-step, water-splitting cycle, based on metal oxides redox pairs bypasses the separation hurdle. Multi-step thermochemical cycles can allow the use of more moderate operating temperatures, but their efficiency is still limited by the irreversibility associated with heat transfer and product separation.

A lower-valence metal oxide can be used for splitting water. A partial reduction of the metal oxide in the absence of a reducing agent would be sufficient for the purpose of splitting water. Hydrogen and oxygen can be derived in different steps, eliminating the need for high temperature gas separation in single-step direct water splitting. Redox pairs such as Fe_3O_4/FeO, ZnO/Zn, TiO_2/TiO_x (with X<2), Mn_3O_4/MnO and Co_3O_4/CoO are used to reduce the reduction temperature.

SOLAR ENERGY

Solar energy has the promise of becoming an important energy alternative. Photovoltaic (PV) cells are becoming less costly and finding more and more applications. Many small and large nations are increasing their use of solar energy.

A typical 100 cm silicon cell produces a maximum current of just under 3 amps at a voltage of around 0.5 volts. Since many PV applications involve charging lead-acid batteries, which have a typical nominal voltage of 12 volts, modules often consist of around 36 individual cells wired in series to ensure that the voltage is usually above 13 volts which is sufficient to charge a 12 volt battery even on overcast days.

When cells are delivering power to electrical loads in real-world conditions, the intensity of solar radiation often varies over time. Many systems use a maximum power point circuit that automatically varies the load seen by the cell in such a way that it is always operating around the maximum power point and so delivering maximum power to the load.

In a typical monocrystalline module, the open circuit voltage is 21 volts and the short circuit current is about 5 amps. The peak power output of the module is 73 watts, achieved when the module is delivering a current of some 4.3 amps at a voltage of 17 volts.

PV Materials

Silicon is only one material suitable for photovoltaics (PV). Another is gallium arsenide (GaAs), a compound semiconductor. GaAs has a crystal structure similar to that of silicon, but consists of alternating gallium and arsenic atoms. It is suitable for use in PV applications since it has a high light absorption coefficient and only a thin layer of material is required.

They can also operate at relatively high temperatures without the performance degradation which silicon and many other semiconductors suffer from. This means that GaAs cells are well suited for concentrating PV systems. Cells made from GaAs are more expensive than silicon cells, because the production process is not so well developed, and gallium and arsenic are not abundant materials.

GaAs cells have often been used when very high efficiency is needed regardless of cost as required in space applications. This was also the case with the Sunraycer, a photovoltaic-powered electric car, which in 1987 won the Pentax World Solar Challenge race for solar-powered vehicles. It traversed the 3000-km from Darwin to Adelaide, Australia at an average day-time speed of 66-km per hour.

In the 1990 race the winning car used monocrystalline silicon cells of the advanced, laser-grooved buried-grid type. The 1993 winner was powered by 20% efficient monocrystalline silicon PV cells, which achieved an average speed of 85-km per hour over the 300-km course.

Solar Energy Conversion

Solar cells presently cannot utilize all the energy of sunlight. Light from the sun has an average temperature of about 6300°K. The Helmholtz ratio of sunlight is about 95%. This means that theoretically it should be possible to convert 95% of the radiant-energy to electricity.

In practice, solar cells may only convert about 10% of the radiant-energy into electricity. This means that 90% of the remaining sunlight is converted into thermal-energy. Solar cells are made of semiconductor material and when the sunlight strikes the semiconductor an electrical potential is created by dislodging electrons. This is due to the impact of the photons. Sunlight is made of photons that contain different amounts of photon-energy at different frequencies. The semiconductor material cannot easily be tuned to convert all types efficiently. This means that some photons will not be converted at all because they have too little photon-energy and some photons will only have a part of their energy converted to electricity because they have too much photon-energy.

Solar cells that have several layers of different semiconductors may be much more efficient. Each layer can be tuned to a specific photon-energy range. Another type of solar cell separates the light into different colors and then each color is converted using a different type of semiconductor. This results in a higher efficiency.

Solar cells may also use lenses or parabolic concentrators to convert more of the radiant energy into thermal-energy. The lens increases the brightness of the light and converts some of the photon-energy into thermal-energy.

A vapor cycle engine could be used to convert this thermal-energy into electricity. This could result in an efficiency of nearly 30% in large installations. One of the advantages with this system is that during the night or on cloudy days fuel can be burned in a separate boiler to operate the system. With the right solar concentrator and engine, efficiencies of over 50% are possible.

A modern gas turbine combined cycle has up to 60% efficiency and uses a much lower temperature of combustion than the sun. The difficulty with solar energy is the path that is required to get the sunlight into the process. Most collection schemes allow radiant-energy to escape. One-way coatings generally only work to keep some of the lower energy rays from escaping but not the higher energy rays.

Solar concentrators contract the solar radiation from a relatively large area onto a small area. A parabolic mirror of four feet in diameter covers an area of 4 pi, or 12.57 square feet (1.17 square meters). This surface area is measured on a plane and is slightly less than the surface area of the curved mirror. If the sun is about 20% down from peak strength, its strength should be about 800 watts per square meter. Then the total amount of energy striking the mirror is almost a 1,000 watts.

Solar Heating Collectors

The most common solar collector is the glass-covered flat plate type. Others include concentrating trough and tube-over-reflector collectors, and parabolic collectors. A flat plate collector uses a flat black absorber plate in a container box. Insulation behind the absorber plate and on the sides of the box reduce heat loss. The glazing may be flat glass, translucent fiberglass or clear plastic, on the sun side of the collector, 1 to 2 inches above the absorber.

If the collector fluid is a liquid, runs of black pipe are attached to the metal absorber. If the fluid is air, the absorber is often textured or uses fins to increase the area that can absorb heat. The absorber may also be a blackened screen material through which the air can flow or an extruded EPDM rubber sheet with parallel water tubes running through it, which is called an absorber mat.

Concentrating collectors include tube-over-reflector types with tubes running in a north-south direction. Each tube absorber is enclosed in a glass pipe. This system can collect sunlight even when the sun is very low and is useful throughout the day and the year. Trough collectors use two mirrored sheets of glass set at right angles to make a channel. The sheets reflect sunlight onto a blackened pipe set about two pipe diameters from the sides of the glass. The absorber can be enclosed in glass pipe, or the entire channel may be covered with a flat glass plate.

Solar Growth

Freiburg, Germany, probably has more solar energy projects than anywhere in Europe. The city even has a guide book of examples ranging from solar-powered parking meters to the solar heated headquarters building of the International Solar Energy Society. The guide book is 65 pages long. Solar technology is incorporated into the roofs and walls of buildings.

Freiburg's success in solar follows from municipal resolutions adopted in the mid-1980s to encourage energy conservation and the use of renewable energy sources including wind, water and biomass, as well as solar.

The sun's output is abundant and free, but it is also diffuse. Its potential as a resource has not always been welcomed. In 1985, the old Central Electricity Generating Board in England stated that large-scale electricity generation from solar power had the disadvantages of high cost, large demands on land area and in the United Kingdom (U.K.) low levels of solar radiation. British Nuclear Fuels, stated that 150 square km of solar panels would be needed to produce as much energy as a typical nuclear power station.

But, the sun's energy can be harnessed in various ways. Buildings can capture and retain the suns warmth using passive solar heating. Solar collectors or panels can be added to buildings to generate power. Solar thermal power plants concentrate the sun's heat to raise steam and drive generators.

During the 1970s oil crises, several were built in the southwestern United States. Five are operated by the Kramer Junction Company (KJC) in California's Mojave Desert. KJC's plants use rows of parabolic trough reflectors covering an area of more than 405 hectares. The troughs reflect the sun's rays onto a network of steel tubes containing a fluid which is heated up to 390°C. The fluid is then pumped through heat exchangers to produce steam for generator turbines with a total output of 150-MW.

Photovoltaics are adaptable and do not need deserts or cloudless skies. Application of PV systems to buildings shows that solar electricity can now be produced without needing any extra land. Arrays of PV modules can be designed into new buildings or added to old ones.

Building integrated PV systems have been installed on the roofs and facades of houses, factories, offices, schools, public buildings and stadiums. The power produced can be on site, stored or fed to the grid. Most systems, except for the very smallest, are connected to local supply networks, and with suitable connections and metering, many owners can sell their surplus current to the utility.

The system on the roof of the home of Jeremy Leggett, a former Greenpeace scientist who runs Solar Century, a company that designs PV systems for buildings, is made of solar tiles. These look similar to regular roof tiles but they generate electricity. Leggett's roof generates about 15% more power than he uses, so he sells the excess to the power company. Each solar roof on an average house over its lifetime prevents about 34 tons of greenhouse gas emissions.

Roof Systems

The energy potential of light falling on buildings is huge. A 1999 report for the U.K. Department of Trade and Industry stated that PV systems installed on all available domestic and non-domestic buildings in the U.K. by 2025 could generate almost as much power as the average consumed in a year.

BP Amoco, an oil company that is one of the world's largest manufacturers of photovoltaic cells, claimed that if every south-facing roof and office wall in the U.K. had solar panels, they could generate more than the UK's total power requirements. In a country like Britain, its cool and wet weather would not be expected to produce enough energy from solar. But most studies indicate that solar energy, particularly if complemented by other renewables, could play a more important role than previously thought.

Several countries have industrious programs for solar power. In Europe, Germany started its 1,000 Roofs Programme in 1990. This was a joint effort by federal and state governments for roof-mounted grid connected PV systems in the 1 to 5-kW range. Installation costs were offset by 70% subsidies and over 2,000 systems were approved. The project has since been increased to 100,000 roofs, which is the equivalent of 300-MW. Italy has a 10,000 PV roof programme and the Dutch government is aiming for 100,000 PV roofs by 2010 and 560,000 by 2020.

The European Commission in 1997 proposed to generate 12% of the European Union's (EU) power from renewable sources by 2010. Along with 40,000-MW from wind farms and 10,000-MW from biomass, there would be 50,0000 PV systems on roofs.

Another EU initiative would export 500,000 PV systems to villages outside Europe. These systems would be used for decentralized electrification in developing countries, while increasing the solar manufacturing industry in Europe.

A poll in *Newsweek* indicated that almost 85% of Americans support more federal investments in solar and wind power. In 2001, almost 75% of the voters in San Francisco supported a $100 million bond for solar on buildings in that city. The Sacramento Municipal Utility District (SMUD) has a waiting list of building owners who want solar on their rooftops. It has installed over 10-MW of systems. In some areas of California, Home Depot is selling complete solar power systems.

Japan is subsidizing 10,000 PV installations on domestic buildings, while the United States has the goal of a million solar roofs by 2010, which include solar heating systems as well as photovoltaics.

BP Solar and several financial institutions recommended a U.K. program of at least 70,000 PV rooftops, a national share of the EU target. The U.K. renewables goal would be 5% of power from renewable sources by 2003, and 10% by 2010.

Photovoltaics may provide a revolution in the supply of electric power. Still, to ensure that new buildings contribute to sustainable development a less cautious bureaucracy, which is less resistant to new ideas and not associated with vested interests, especially in the non-renewable electricity industry.

Solar PV is believed to be on the edge of a trillion dollar market. Many oil companies are diversifying into renewables with optimistic expectations. Shell now manufactures PV cells in Germany and the Netherlands and predicts by 2050, that half the world could be powered by renewable energy. BP Amoco and Shell have been installing PV cells in some of its filling stations. BP Amoco believes that by 2050 all Europe's power could be met by solar energy.

The key is a reduction of costs. Solar panels are expensive since photovoltaic technology is still in its infancy. Although the price of PV cells has fallen significantly, PV electricity is still not without a subsidy. As more PV systems are built and installed, the market should result in solar electricity becoming more and more competitive.

There are more than two billion people without access to electricity, according to the United Nations Development Program. When night falls in the developing world, 70% of the population are without electrical lighting. Most of these rural areas are too isolated to be connected to a utility grid.

Solar Electric Light Fund (SELF) is a non-profit charitable organization started in 1990. SELF promotes and develops energy self-sufficiency in developing countries. Using the latest photovoltaic (PV) technology, it brings power to the developing world. Some of SELF's projects include a rural solar project in Karnataka, India, PV systems for up to 10,000 houses in Zimbabwe and equipping rural schools in Southern Africa with solar-powered computers and wireless Internet access.

In the Taos area of New Mexico, the earthship concept recycles water and waste and generates power from PV cells. Two earthship colonies have been started in the foothills of the nearby Sangre De Cristo mountains and designs exist for complete cities of earthship buildings.

Logistical problems have damped solar energy growth. This includes the difficulty of finding reputable contractors to install solar panels.

Since 1998, the California Energy Commission has been pushing a program to encourage homeowners to erect photovoltaic panels on their roofs, offering to subsidize about one-third of the cost.

When the 2000 California energy crisis struck, with its power interruptions and steep rate increases, interest spiked. One solar contractor's web site receives 3,000 hits a day. The Los Angeles power department also reports a surge in interest since the energy crisis began. Customer demand is reported to be very high according to the Department of Water and Power. They receive about 1,000 calls on some days about the solar subsidy program.

In 2000, Los Angeles, California announced a goal of 100,000 roofs covered with solar electric panels by the end of the decade. The Los Angeles Department of Water and Power began offering subsidies that would reimburse buyers for half the price of each new solar energy system. For an average home, a photovoltaic package may cost between $10,000 and $20,000, including installation before the rebate. In Los Angeles, the city's only solar panel manufacturer has not been able to supply enough systems to meet demand. The installation of a one-kilowatt solar electric system on a home in the San Fernando Valley was the first to be awarded a rebate by the Los Angeles power department in March, 2001.

The price of photovoltaics continues to drop and interest is continues to grow. States such as New York, Arizona, Florida and Washington have joined California in a major effort to allow homes and businesses to use solar power.

The systems are almost maintenance free, but panels must be cleaned of dirt, dust and leaves. They need to be installed on roofs without shade on south-facing roofs.

The reliability and cost of solar electric technologies should continue to improve, although solar power only accounts for less than 1% of all power consumed. The U.S. produces about 300 megawatts of electricity with solar which is about the same amount produced by a mid-size traditional power plant. If solar energy is to provide a significant part of the world's energy needs, the cost of solar must be competitive with other energy sources such as natural gas, nuclear or coal.

California has nine solar stations with 11 square miles of mirrors focused on steam drums that drive steam turbines. They can generate 413 megawatts (MW) of electricity which is less than 1% of the state's capacity. Because the sun sets at night and is sometimes attenuated by clouds, these plants produce only 0.3% of California's electricity.

They are supported by federal solar power tax credits along with California's Public Utilities Regulatory Policy Act (PURPA) contracts and renewable power subsidies. When these tax credits were interrupted for eleven months in 1991, the plants' operator, LUZ, immediately went bankrupt. Today SEGS, an Israeli government corporation, operates them at a loss.

The current trend for photovoltaics is not to erect large centralized solar farms in the desert, which started in the 1980s, but utilize distributed generation with individual units on rooftops.

One problem has been the cost of the solar panels. Los Angeles began its solar program after state legislators mandated that utilities spend about 3% of their revenue on efficiency, conservation and renewable energy. For solar, the power department had $75 million to spend over the next five years.

The power department will pay $5 for each watt of solar installed on a residence or business. Homeowners typically purchase a 1- or 2-kW system meaning that the municipal utility pays between $5,000 and $10,000 of the cost.

The systems that are eligible for rebates must be tied into a utility's electric power grid. If there is a surplus of solar power, it goes back into the power grid.

A 2-kilowatt solar system can supply an average-sized home with 20 to 80% of its electrical needs, depending on how many lights, appliances and air conditioners are running, and how efficient they are.

After the subsidy, and depending on how the system is paid for (cash or borrowed money), a solar system may pay for itself in as little as 6 years and as much as 36 years.

The L.A. Department of Water and Power was not deregulated along with the three major utilities in California. It has some of the lowest power rates in the state making the economical argument harder to make.

To receive the full $5 per watt subsidy, the L.A. Department of Water and Power requires a homeowner to purchase solar panels from a manufacturer based in the city. This was done to encourage the local growth of an emerging industry. However, there were no solar panel makers located in Los Angeles. After some negotiations, Siemens Solar Industries, based in Camarillo, California decided to open a solar panel manufacturing plant in Los Angeles. But, it has not been a complete facility. The plant does some final assembly and then the units are returned to Camarillo for final testing and shipping.

Siemens Solar reports the interest from consumers is sometimes overwhelming and that supply has been a problem. Most U.S. manufactured units are shipped to countries such as Germany, Japan and Scandinavia, which have had generous subsidies for years.

References

Braun, Harry, *The Phoenix Project: An Energy Transition to Renewable Resources*, Research Analysts: Phoenix, AZ, 1990.

Cothran, Helen, Book Editor, *Global Resources: Opposing Viewpoints*, Greenhaven Press,: San Diego, CA, 2003.

Romm, Joseph J., *The Hype About Hydrogen*, Island Press: Washington, Covelo, London, 2004.

CHAPTER 9

INFRASTRUCTURE CHOICES AND NUCLEAR HYDROGEN

Hydrogen infrastructure will depend on where the hydrogen is produced and what form it is stored. The major choices are onboard hydrogen production, centralized production, and production at fueling stations.

Reforming either methanol or gasoline into hydrogen onboard a vehicle is likely to be less efficient than stationary reforming. Onboard reformers produce less pure hydrogen, which reduces the fuel cell's efficiency. The overall efficiency for gasoline and methanol fuel cell vehicles is likely to be much lower than for hydrogen fuel cell vehicles. The onboard reforming of gasoline to hydrogen may produce some modest emission benefits.

Onboard gasoline reforming could serve as an interim step and accelerate the commercialization of PEM fuel cells. It does not require a hydrogen infrastructure.

Onboard methanol reformers are likely to be even less efficient than gasoline reformers. For the immediate future, increases in methanol production are likely to come from overseas natural gas.

The existing methanol infrastructure could handle some methanol fuel cell cars at a low cost of about $50 per car. But any significant use of methanol as a transportation fuel would require additional investments in fuel production and delivery, which may be somewhat less than a hydrogen infrastructure.

Hydrogen fuel cell vehicles should become much more efficient and popular so any investments in methanol infrastructure could become lost. The fuel infrastructure might have to be changed from gasoline to methanol and then from methanol to hydrogen. An affordable direct methanol fuel cell is needed as well as an affordable way to generate large quantities of methanol from renewable sources.

Fuel companies like Royal Dutch/Shell have invested heavily in hydrogen. Transition fuels such as onboard methanol-to-hydrogen conversion would require infrastructure investments, which would be difficult to justify.

CENTRALIZED PRODUCTION

Centralized production of hydrogen should provide less expensive hydrogen than production at local fueling stations. Resource centered hydrogen production near large energy resources, such as sources of natural gas are not carbon free like wind power and biomass. Centralized units can time their electricity consumption for compression during off-peak rates compared to local fueling stations.

Hydrogen delivery with tanker trucks carrying liquefied hydrogen, is energy-intensive. Pipelines are less energy intensive option, but they are expensive investments. Until there are high rates of utilization the high capital costs hold back investment in these delivery systems. Trucks carrying compressed hydrogen canisters may be used for the initial introduction of hydrogen.

One study by Argonne National Laboratory estimates infrastructure costs for fueling about 40% of the vehicles on the road with hydrogen would be $600 billion. Production of hydrogen at local fueling stations is promoted by those who want to deploy hydrogen vehicles quickly. The hydrogen could be generated from small stream methane reformers. Electrolysis is considered to be more expensive.

Fueling stations would have a reformer, hydrogen purification unit and multi-stage hydrogen compressor for high-pressure tanks. There would also be a mechanical fueling system and on-site high-pressure storage. Advances will be required in reformers and electrolyzers, compressors, and systems integration.

The National Renewable Energy Laboratory, found that forecourt hydrogen production at fueling stations by electrolysis from grid power was most expensive, at $12/kg with forecourt natural gas production at $4.40/kg.

Natural gas may be the wrong fuel on which to base a hydrogen-based transportation system. A large fraction of new U.S. natural gas consumption will probably need to be supplied from overseas. While these sources are more secure than sources for oil, replacing one import with another does not move us towards independence.

Natural gas can be used far more efficiently to generate electricity or to cogenerate electricity and steam than it can be to generate hydrogen for use in cars. Using natural gas to generate significant quantities of hydrogen for transportation would, for the foreseeable future, damage efforts to battle CO_2 emissions.

THE FUTURE

One scenario by Shell predicts natural gas consumption increasing through 2025 and then dropping due to supply problems. Renewable energy grows and by 2020 a variety of renewable sources supplies a fifth of the power in many developed countries. By 2025 biotechnology, materials advances and sophisticated power grid controls allow a new generation of renewable technologies to spread. Advances in storage technology may also add renewables. Oil becomes scarce by 2040, but highly efficient vehicles using liquid biofuels from biomass farms solve this problem with help from super clean diesel fuel made from natural gas. By 2050 renewables grow to a third of the world's primary energy and supply most incremental energy.

These are dramatic increases in renewable energy and energy efficiency. In the U.S. these renewables, such as wind and solar power, now represent less than 1% of electric power generation.

From 1975 to 2000, the world gross domestic product (GDP) more than doubled while primary energy use grew by almost 60%. From 2025 to 2050 in Shell scenario, the GDP nearly doubles, but primary energy use grows by only 30%. This means that energy use would have to be twice as efficient.

Another scenario by Shell sees a technology revolution based on hydrogen. It is based on development of fuel in a bottle for fuel cell vehicles. Two liter bottles would be used to hold enough fuel to drive forty miles. Fuel bottles would be distributed like bottled water through existing distribution channels including vending machines. Consumers could get their fuel anywhere and at any time.

In this scenario, by 2025, one-quarter of the fleet vehicles use fuel cells, which make up half of new sales. Renewables grow quickly after 2025.

About one billion metric tons of CO_2 are sequestered in 2025. After 2025, hydrogen is produced from coal, oil and gas fields, with carbon dioxide extracted and sequestered cheaply at the source.

Large-scale renewable and nuclear energy techniques for producing hydrogen by electrolysis come online by 2030. Global energy use almost triples from 2000 to 2050. Global nuclear power production also nearly triples during this period. Natural gas use is large in this scenario, and its use more than triples over these 50 years. Renewable energy is also abundant.

By 2050, CO_2 sequestration is over 8 billion metric tons per year, one-fifth of emissions. The world is sequestering more CO_2 than the United States produces now from coal, oil and natural gas use.

Shell maintains that these are not predictions but conception exercises.

The fuel in a bottle would have to be like liquid propane distribution to-day, but propane is liquid at a much higher temperature and lower pressure than hydrogen. The form of hydrogen contained could not be high-pressure storage, since that would be too bulky, too heavy, and almost certainly too dangerous to distribute by vending machine. It could not be metal hybrids, since that would be even heavier. If it were liquid hydrogen, since that could not be dispensed in small, portable, lightweight bottles with today's technology. None of these would seem to be easily used by the consumer to fuel a hydrogen vehicle.

The Shell concept studies imply that fuel cell sales start with stationary applications to businesses that are willing to pay a premium to ensure highly reliable power without utility voltage fluctuations or outages. This demand helps to drive fuel cell system costs below $500 per kW, providing the stage for transportation uses which incite additional cost drops to $50 per kilowatt in the next half decade.

But, the high-reliability power market may not drive demand and cost reductions, especially for proton-exchange membrane (PEM) fuel cells. This scenario leaves many questions in its assertions and assumptions.

Shell estimates that the cost in the U.S. to supply 2% of cars with hydrogen by 2020 is about $20 billion. This suggests that serving 25% of cars in 2025 would be very costly.

By 2025 the world is sequestering 1 billion metric tons of CO_2 per year while simultaneously producing hydrogen and shipping it hundreds of miles for use in cars. This is equivalent to sequestering the CO_2 produced by more than 700 medium-sized generation units, about two-thirds of all coal-fired plants in the United States today.

The U.S. Department of Energy (DOE) has announced the billion-dollar, FutureGen project to design, build, construct and demonstrate a 275-megawatt prototype plant that cogenerates electricity and hydrogen and sequesters 90% of the CO_2. The goal of this project is to validate advanced coal near zero emission technologies that by 2020 could produce electric power that is only 10% more costly than current coal generated power. This would be an advanced system that in five years Shell suggests would grow to 700 worldwide.

Advances can occur quickly in technology, these would be needed in hydrogen production and storage, fuel cells, solar energy, biofuel production and sequestration. Government and industry would need to spend hundreds of billions of dollars to bring these technologies to the marketplace. Political obstacles to tripling nuclear power production would need to be set aside. Natural gas supplies would need to be increased.

Shell has made extensive investments in renewable energy and hydrogen and has been a leader in reducing greenhouse gas emissions.

One problem is cost-effectiveness, hydrogen must be able to compete with alternative strategies including more fuel-efficient internal combustion engine vehicles. In the near term, hydrogen is likely to be made from fossil fuel sources. The annual operating costs of fuel cell power are likely to be higher than those of the competition in the foreseeable future.

Competition

Underestimating the competition may be the single biggest reason why newer cleaner energy technologies achieve success in the market more slowly than expected. Renewable energy has been the focus of public and private research and development (R&D) in the past few decades.

In general, renewable technologies have met expectations with respect to market penetration. Wind has met projections from the 1980s, although earlier projections were overly optimistic. In biomass applications, market penetration has exceeded previous projections.

There are those who believe that global warming is the most potentially catastrophic environmental problem facing the nation and the planet this century and it is therefore the problem that requires the most urgent action on the part of government and the private sector. They may advocate that spending money on building a hydrogen infrastructure would take away resources from more cost-effective measures. A hydrogen infrastructure may be critical in achieving a major CO_2 reduction in this century.

In the first half of the 21st century, alternative fuels could achieve greater emissions and gas savings at lower cost, reducing emissions in electricity generation. This is true for natural gas as well as renewable power in the near future.

A natural gas fuel cell vehicle running on hydrogen produced from natural gas may have little or no net CO_2 benefits compared to hybrid vehicles. Natural gas does have major benefits when used to replace coal plants. Coal plants have much lower efficiencies at around 30% compared to natural gas plants at 55%. Compared with natural gas, coal has nearly twice the CO_2 emissions, while gasoline has about one third more CO_2 emissions than natural gas.

In the United States, vehicle emissions other than CO_2, have been declining steadily. Noxious emissions are being reduced by federal and state regulations and the turnover of the vehicle fleet. As the older vehicles go out of service, they are replaced with newer and cleaner vehicles. The federal Clean Air Act Amendments of 1990 seem to be working. In the 1990s, Tier 1 standards greatly reduced tailpipe emissions of new light-duty vehicles which includes cars and most sport utility vehicles.

By 2010, Tier 2 standards should further reduce vehicle emissions by extending regulations to larger SUVs and passenger vans. The use of gasoline with a lower sulfur content will also reduce emissions and it also makes it easier to build cars that achieve further reductions.

These standards should allow new U.S. cars to be extremely free of air pollutants. But, the Clean Air Act does not cover vehicle CO_2 emissions. Many new cars are called near zero emissions vehicles by their manufacturers and may have tailpipe emissions cleaner than some urban air. Hydrogen fuel cell vehicles will have almost no emissions besides some water vapor and would be much cleaner.

The U.S. has been building new natural gas power plants because they are more efficient and cleaner. By 2003, the nation had more than 800 gigawatts (GW) of central station power generation. One gigawatt is 1,000 megawatts (MW) and is about the size of a very large existing power plant or three of the newer, smaller plants. Almost 145 gigawatts were added from 1999 to 2002 and almost 96% of this was natural gas. This included 72 gigawatts of combined-cycle power and 66 gigawatts of combustion turbine power which are used generally when demand is high.

The Energy Information Administration predicts an increase in coal generated power. The EIA estimates that from 2001 to 2025, about 75-GW of new coal plants will be built. Over 90% of the coal plants are projected to be built from 2010 to 2025.

The EIA forecast also predicts that existing coal plants will be used more often. From 2001 to 2025, the EIA estimates a 40% increase in coal consumption for power generation. This could increase U.S. greenhouse gas emissions by 10%.

The rising demand for natural gas already affects North American supplies and has pushed up prices. Canada is an important source of our imported natural gas, but it has little capacity left to expand its production.

While not as energy-intensive a process as liquefying hydrogen, cooling natural gas to a temperature of about -260°F and transporting the resulting liquid has an energy penalty of up to 15%, according to the Australian Greenhouse Office. From a global standpoint, it might be better to use foreign natural gas to offset foreign coal combustion than to import it into the United States in order to turn it into hydrogen to offset domestic gasoline consumption. The projected growth in global coal consumption could be an even bigger CO_2 gas problem than the projected growth in U.S. coal consumption.

By 1999, there were over 1,000-GW of coal power generating capacity around the world. About one third of this is in the United States. From 2000 to 2030, more than 1,400-GW of new coal capacity may be built, according to the International Energy Agency of which 400-GW will be used to replace older plants.

Carbon Capture

These plants would need to use carbon capture equipment or their estimated carbon emissions could equal the fossil fuel emissions from the past 250 years. Carbon capture and storage (CCS) is an important research area but widespread commercial use may be years away.

Many of these plants may be built before CCS is ready and we will need to use our electricity more efficiently to slow the demand for such power plants, while building as many cleaner power plants as possible. Natural gas is far more cleaner for this power than coal. Generating hydrogen with renewables may be needed in order to avoid building coal-fired plants. More electricity from renewable power would reduce the pressure on the natural gas supply and reduce prices. The United States could have essentially carbon-free electricity before 2050 with hydrogen fuel playing a key role.

Some studies indicate that higher carbon savings can be achieved by displacing electricity from fossil fuel power stations. Abundant renewable power and the practical elimination of CO_2 emissions from electricity generation could take 30 years. The United Kingdom's power generation mix has less CO_2 emitted per megawatt-hour by one third compared to United States. The U.K. has moved away from extensive coal power generation in the past few decades and is aggressively pushing renewable energy and cogeneration.

The gasoline tax in Europe has some effect on auto use and fuel consumption. The United Kingdom, France and Germany have gasoline taxes of more than $2.00 per gallon, which is about five times the federal gasoline tax in the U.S.

Fuel Efficiency

Fuel efficiency was used successfully in the late 1970s and early 1980s. A 2002 report by the National Academy of Sciences stated that automobile fuel economy could be increased by 12% for small cars and up to 42% for large SUVs. The study did not include the greater use of diesels and hybrids. Other studies have indicated that even greater savings are possible while maintaining or increasing passenger safety.

Europe has an agreement with automakers that would reduce the CO_2 emitted per mile by 25% from 1996 to 2008 for light duty vehicles, which would result in an average fuel efficiency of about 40 mpg. Japan has a similar goal.

CALIFORNIA STANDARDS

California has the strictest air quality standards in the U.S. and has been promoting zero emission vehicles (ZEVs). Electric cars have not been a success and fuel cell vehicles are viewed a logical step to ZEVs. In California most of the electricity comes from natural gas, nuclear, renewables other low carbon generating sources, resulting in about half the CO_2 emissions per kilowatt-hour compared to total U.S. emissions. One of the CAFCP's goals is to demonstrate advanced vehicle technology by operating and testing vehicles under real-world conditions in California. Other goals are to demonstrate the viability of alternative fuel infrastructures including hydrogen and methanol stations. This will allow the state to explore a path to commercialization and identify potential problems while increasing public awareness on fuel cell vehicles.

The first three dozen fuel cell vehicles on the road in California include buses and light-duty vehicles with dozens more scheduled to be added later. Eight hydrogen fueling stations are planned to dot the state from Los Angeles to Sacramento. California plans to establish a hydrogen freeway that would provide fueling stations for hydrogen-powered vehicles at convenient locations across the state. But, a state cannot build a nationwide fueling infrastructure. California can play a leadership role in fuel cell vehicles in the United States, much as Iceland is doing for the rest of the world.

NUCLEAR POWER

Nuclear power plants could be a major source of hydrogen. Electrolysis of water with electricity from a nuclear power plant may not be economical now but the waste heat from these plants may be high enough to generate hydrogen by thermochemical decomposition of water into hydrogen and oxygen.

Thermochemical water splitting at temperatures above 750°C could provide a 40 to 50% efficiency in hydrogen production. Cogeneration of electricity might raise the overall efficiency to 60%.

This approach would have to compete with other emerging hydrogen generation technologies. The DOE is currently investigating thermochemical hydrogen production with nuclear power. Their goal is a demonstration of commercial production by 2015.

Nuclear generated hydrogen could be a practical solution, in the U.S. About 100 nuclear water splitting plants could replace a vital portion of U.S. transportation fuel with hydrogen.

In a deregulated market environment, nuclear power is not yet cost competitive with coal and natural gas. There are also concerns about safety, environmental health and terrorism. Nuclear power also has not resolved all problems in long-term management of radioactive wastes and needs to be shown as a safe and economical source of hydrogen before attracting the investment capital to build 100 new plants.

The hydrogen would probably be generated from nuclear power plants away from urban areas so there would be infrastructure costs for delivering the hydrogen. Additional nuclear power capacity would be needed to meet the needs of a hydrogen economy. The nuclear waste issue to still being resolved and modular plants similar to those used in France could greatly improve safety and licensing issues.

Nuclear fission power plants were initially thought to be the answer to diminishing fossil fuels. Although the enriched uranium fuel was also limited, a more advanced nuclear reactor called the breeder would be able to produce more radioactive fuel, in the form of plutonium, than consumed. This would make plutonium fuel renewable. Although plutonium has been called one of the most toxic elements known, it is similar to other radioactive materials and requires careful handling and can remain radioactive for thousands of years.

Conventional nuclear reactors and advanced breeder reactors were America's primary energy strategy since the 1950s to resolve the fossil fuel problem.

When a reactor accident occurred in 1979 at Three Mile Island in Pennsylvania, public and investor confidence in nuclear fission were cracked. The accident was triggered by the failure of a feedwater pump that supplied water to the steam generators. Backup feedwater pumps were not connected to the system as required, which caused the reactor to heat up. The safety valve then failed to act which allowed a radioactive water and gas leak.

Radioactivity

As uranium undergoes fission, the uranium atoms split and release neutrons. Some of these neutrons split other uranium atoms, which produce radioactive waste products. The net result of the fission process is the generation of intense heat which is used to generate steam for turning the generators.

A nuclear reactor and a nuclear weapon both release a number of neutrons during the fission process over a given period of time. If the number of neutrons are limited for triggering the fission chain-reaction, the reaction can be controlled for producing energy. If too many neutrons are released, the chain-reaction will accelerate, resulting in an explosion. To prevent this from happening, nuclear reactors use control rods and water circulation to regulate the fission process by absorbing the extra neutrons.

However, some of these neutrons will pass into the steel structures which hold the fuel assemblies and the cooling water which flows between them. Other neutrons may penetrate the concrete shielding outside the steel reactor vessel. These neutrons are absorbed by the atoms of iron, nickel and other elements that they pass through.

When atoms absorb neutrons, they become unstable and release particles making them radioactive for differing lengths of time. A material like nickel-59 has a half-life of 80,000 years, it needs to be shielded for about a million years.

It is highly unlikely that a nuclear fission power plant would ever explode like a nuclear bomb, but a loss of coolant accident could result in a melt down condition.

In a melt down, a large amount of radiation could be released at ground-level. A nuclear or conventional chemical or steam explosion could disperse much of the radioactive particles into the atmosphere. This is essentially what happened when the Chernobyl accident occurred in the Soviet Union on April 26, 1986.

Several features made the accident unique to a Soviet style reactor. One was the use of graphite as a moderator, which caught fire. Another was the absence of water to contain radioactivity. But, the most important may have been an inadequate containment structure. There were also problems in controlling the stability of the reactor and the control rods had to be changed frequently in order to keep the reactor stable.

Before the accident at least 6 safety mechanisms were disconnected to conduct experiments to increase the output of the reactor. This was the direct cause of the accident and as the power output surged from 7 to 50% in a few seconds there was a loss of coolant. The heat then melted the graphite rods used as a moderator.

An experiment to find out how long power was generated as the reactor unit was shut down was authorized. But, automatic shut-down mechanisms were blocked that may have come into operation at low capacity levels. These included the reactor's emergency cooling system and its low water level safeguard. Extra pumps were also turned to raise the amount of steam going to the generator. These pumps were over the allowable level.

The Soviet government initially assured those living in the area of the Chernobyl accident that the impact would not be significant. But, three years after the accident cancer rates doubled among residents of contaminated regions. Moscow News reported that more than half of the children in the Narodidiehsky region of the Ukraine have illnesses of the thyroid gland, which exposure to radiation can cause. Soviet officials admitted that they greatly underestimated the health problems caused by the reactor explosion and fire.

Military Reactors

The U.S. Navy has an excellent performance record with its fleet of nuclear surface ships and submarines. There are major differences in the size of the nuclear systems used by the U.S. Navy. The Nautilus submarine used 60 megawatt reactor which was scaled up to 600, 900 and then over 1000 megawatts for commercial power plants.

The reactors used by the U.S. Navy were initially about six times more costly per kW than commercial units. In 1973, it cost about $2,400 per kW to build a U.S. Navy nuclear reactor, compared to $400 per kW for commercial plants at that time. By the 1990s capital costs for commercial reactors would be reaching $3,000 per kW.

Advanced Reactors

In the 1980s, a new generation of nuclear fission reactors called Integral Fast Reactors was under development by the U.S. Department of Energy. This was a liquid sodium cooled reactor which was expected to be safer with minimal corrosion. It was also to be more efficient and able to use 15 to 20% of the uranium fuel instead of 1 to 2%.

The breeder reactor creates more fuel than it consumes by converting U-238 into plutonium 239. It uses a 10-20% enriched core of uranium and plutonium surrounded by a shell of U-238. The neutrons emitted by the core are absorbed by the U-238 in the shell which is transmuted into plutonium-239. Less radioactive waste was also a feature compared to the light water reactors in use in the U.S.

The Integral Fast Reactor would also be capable of breeding plutonium which could be used as nuclear fuel. This type of reactor was seen as the key to a nuclear future. Liquid sodium is a volatile substance that can burst into flames if it comes into contact with either air or water. An early liquid sodium-cooled breeder reactor, the Fermi I, had a melting accident when 2% of the core melted after a few days of operation. Four years later when the reactor was about to be put into operation again a small liquid sodium explosion occurred in the piping.

France has the largest implementation of breeder reactors with its Phenix reactor of 250-MW and Super-Phenix with 1200-MW. The Phenix was put into operation in 1973 and the Super-Phenix in 1984. Japan has its 300-MW Monju reactor which was put into service in 1994. While India has the 500-MW PFBR and 13.2-MW FBTR.

These reactors produce about 20% more fuel than they consume. Optimum breeding allows about 75% of the energy in natural uranium to be used compared to 1% in a conventional light water reactor.

Nuclear Fusion

Nuclear fusion reactors do not split uranium atoms. They fuse hydrogen atoms in a process similar to that which occurs in the Sun and other stars. Although fusion physics is a common occurrence in stars, controlled fusion experiments continue. In 1994, the Tokamak facility at Princeton reached a fusion plasma temperature of 510 million degrees and had a power output of 10.7 megawatts.

The basic fuel in a fusion reactor is deuterium, a heavy form of hydrogen found in water. One out of every 6,500 molecules of ordinary water contains deuterium. It cost about 10 cents to separate the deuterium from a gallon of ordinary water. One teaspoon of deuterium has the energy equivalent of 300 gallons of gasoline and 1,000 pounds of deuterium could operate a 1,000 megawatt power station for a year.

The waste from fusion is much less toxic that of fission reactors. Most of the waste will occur in the surrounding materials of the process, the steel vessels and piping. The materials have half-lives in tens rather than thousands of years and are expected to be reusable in 20 years.

Much research has been done on this technology but instability and efficiency problems remain. The U.S., Japan, France, Germany, the Soviet Union and other European countries have all conducted fusion research. Some fusion energy systems are expected to use energy pellets which would make them similar to coal-fueled power plants.

Energy production in the future could be greatly altered with small, clustered, safe high-temperature fusion reactors burning cheap, abundant fuel.

Cold Fusion

In 1989, two scientists, at the University of Southampton in England, held a press conference to announce that they had succeeded in generating a fusion reaction that produced more energy than the reaction consumed at room temperature. They felt that commercial reactors based on the new low-temperature fusion process could be in operation in about 20 years. However, many experts were skeptical of their claims and they pointed out that the announcement occurred at a press conference rather than from a paper at a technical conference. It is impossible to know if the cold fusion process is valid. The possibility of such a breakthrough in nuclear energy could have a profound impact on global energy. Although, there is still the issue of radioactive wastes that will be generated from such nuclear reactions.

Nuclear Economics

The cost of nuclear power has been aided by government support of the research and development in the United States. The government also covers those costs not met by the utility for waste disposal and decommissioning.

Nuclear operating costs do not include the construction and operation of the U.S. government uranium fuel enrichment facilities. When all three of these enrichment facilities were operating at full capacity, their power consumption was similar to that of the country of Australia. Other excluded operating costs include Federal regulation, long term waste disposal and any health costs that are associated with people being exposed to radiation.

Decommissioning

Decommissioning is used to describe closing down a nuclear reactor once its useful life has been completed. This is the aftermath of the nuclear fuel cycle. Utilities and nuclear waste processing companies have no long-term legal or financial responsibility to manage the radioactive wastes.

In 1963 a $32 million West Valley, New York reprocessing plant was opened near Buffalo and began to reprocess nuclear wastes in 1966. It only operated for 6 years before its operator, Nuclear Fuel Services (NFS), a subsidiary of W.R. Grace's Davison Chemical Company, discarded the facility. There were 2 million cubic feet of radioactive material left behind along with 600,000 gallons of radioactive liquid waste that was seeping into the Cattaraugus Creek, which flows into Lake Erie, from which the city of Buffalo obtained its drinking water. The cost of cleanup was estimated to be $1 billion.

Nuclear Power Accidents

Highly publicized nuclear accidents such as those that occurred at Chernobyl and Three Mile Island are considered anomalies. Nuclear power plants have many safety measures in place to prevent radiation leaks. The small amount of radioactive waste produced by nuclear reactors is controlled and usually contained in the plant facility.

Nuclear is an energy option that provides about 20% of our power. France uses nuclear energy to generate almost 80% of its electric power and a number of other countries are more dependent on this energy option than the United States although the technology was invented and developed here. Nuclear power could make the U.S. less dependent on foreign oil.

Safety Issues

The nuclear power industry has been at a standstill in the United States based on fears that nuclear is too dangerous. Besides France at 80%, Belgium generates 60% of its power from nuclear, Switzerland 42%, Sweden 39%, Spain 37%, Japan 34% and U.K. 22%. These countries that generate a higher percentage of their power with nuclear energy than the U.S. have done so without any loss of life or harm to the environment.

No form of power generation is 100% safe but nuclear power may be safer than many other alternatives for generating large amounts of electrical energy, such as oil and coal plants. This is because the fuel in a nuclear power plant is highly concentrated. One uranium fuel pellet which measures about 0.3-inch diameter by 0.5-inch long can produce the equivalent energy of 17,000 cubic feet of natural gas, 1,780 pounds of coal, or 149 gallons of oil. Since relatively little fuel is used, relatively little waste is produced and this waste is contained within the plant walls. This is not the case with fossil fuel plants, which emit tons of pollutants into the atmosphere. Some nuclear power plants have cooling towers that emit water vapor.

In 2000 an issue of *Foreign Affairs* concluded that fossil fuel electrical power plants are more hazardous to humans than nuclear power plants. *Foreign Affairs* is the journal of the Council on Foreign Relations (CFR).

The article stated that pollutants from coal burning plants cause an estimated 15,000 premature deaths in the United States every year and that a 1,000-megawatt (MW) coal-fired power plant releases about 100 times as much radioactivity into the environment as a comparable nuclear plant.

A 1,000-MWe power plant uses 2,000 railroad cars of coal or 10 supertankers of oil but only 12 cubic meters of natural uranium every year. Fossil fuel plants can produce thousands of tons of noxious gases, particulates, and heavy metal bearing radioactive ash along with solid hazardous waste. There are up to 500,000 tons of sulfur from coal, more than 300,000 tons from oil, and 200,000 tons from natural gas. A 1,000 MWe nuclear plant releases no noxious gases or other pollutants and much less radioactivity per capita than is encountered from airline travel, a home smoke detector, or a television set.

Nuclear Risks

Nuclear power reactors have the following safety issues: nuclear waste, plutonium buildup and radioactivity. Nuclear waste from the reactor fuel consists of uranium that has been formed into a usable metal alloy and provided as small pellets, rods, or plates. The fuel is encapsulated with a metal cladding, such as zircaloy, which adds mechanical strength and also prevents radioactive contamination.

Nuclear reactor waste or spent nuclear fuel consists of the fuel pellets that have been used in a reactor over a period of time, usually about 3 years, and have lost their ability to provide enough energy. This spent fuel is still radioactive and must be shielded to prevent any release.

Spent fuel is stored in shielded basins of water or dry vaults. As the radioactive decay drops to safe levels, it may take hundreds to thousands of years. The nuclear waste containers are designed for an underground storage period of at least 10,000 years.

Spent fuel will be stored on a permanent basis once a national repository is approved. The planned nuclear waste facility at Yucca Mountain is still involved in ongoing environmental impact studies. The opening of a national long-term storage site is over 12 years behind schedule because of opposition.

Some countries, such as France, have progressive nuclear fuel recycling programs where a large percentage of the unused uranium and the small amount of plutonium produced in the spent fuel is salvaged and then processed into new reactor fuel. According to the Nuclear Energy Institute (NEI), only 3% of spent fuel is waste another 96% is unused uranium and 1% is unused plutonium created during the fuel cycle.

Nuclear fuel recycling allows more efficient nuclear fuel usage and less buildup of nuclear waste. Nuclear power reactors are designed to minimize plutonium build up and much of the plutonium that is produced inside the reactor is used during an ordinary fuel cycle.

The amount of radiation that is emitted by nuclear power plants, with their thick shielding is quite low. Environmental Protection Agency (EPA) guidelines limit the annual whole body dose to 25 millirems for uranium fuel operations. According to the National Council on Radiation Protection and Measurements (NCRP), and the EPA, the natural background radiation from the Earth's crust can be 23 millirems per year at the Atlantic Coast and 90 millirems per year on the Colorado Plateau. Radiation inside the human body is about 40 millirems per year from the food and water we take in and can be up to 200 millirems per year from radon in the air. The annual radiation dose from outer space can be 26 millirems at sea level or 53 millirems at elevations of 7,000 to 8,000 feet. The dose from a medical X-ray is about 20 millirems, and the dose from a 1,000-mile airline flight is about 1 millirem. So a cross country air trip and return can add up to over 5 millirems. We can also receive 1-2 millirems annually from watching television or using computers and can get another 7 millirems each year from living in a brick building. We could receive .03 millirem annually by living 50 miles away from a coal-fired power plant, but only .009 millirem by living 50 miles away from a nuclear power plant.

Radioactivity, radioactive elements and nuclear reactors exist in nature. At least 14 natural fission reactors have been found in the Oklo-Okelobondo natural uranium formation in Gabon on the west coast of Africa. These fossil reactors had sufficient amounts of U-235 to allow chain reactions to occur.

Nuclear power plants use multiple layers of protection from the radioactive particles inside the reactor core. A serious accident can cause the release of radioactive material into the environment. It is not a nuclear explosion, because the uranium fuel used in a nuclear power plant does not contain a high enough concentration of U-235.

For an explosion to occur, the uranium fuel inside the reactor would have to be enriched to about 90% U-235, but it is only enriched to about 3.5%.

The worse nuclear power plant disaster on record occurred when the Chernobyl reactor in the Ukraine had a hot gas explosion. If this occurred in a Western nuclear power plant, the explosion would have been contained because Western plants are required to have a containment building. This a solid dome of steel reinforced concrete that contains the reactor. The Chernobyl plant did not have this containment feature, so the explosion blew through the roof of the reactor building allowing radiation and reactor core parts to escape into the air.

The design of the Chernobyl plant was flawed in other ways as well. Western reactors are designed when operating to maintain negative power coefficients of reactivity that prevent such runaway accidents. The Chernobyl plant would not have been issued a license to operate in the U.S. or other Western countries. The Chernobyl accident was in many ways the worse possible scenario having an exposed reactor core and roofless building. Thirty-one plant workers and firemen died directly from the radiation exposure and it is projected that over 3,400 local residents will eventually acquire and die of cancer due to radioactive exposure.

The damage was much greater when 2,300 were killed and as many as 200,000 others injured in a few hours when a toxic gas cloud escaped from the Union Carbide pesticide plant in Bhopal, India.

The worst nuclear power accident in the U.S. occurred at the Three Mile Island plant in Pennsylvania. In this accident no one was killed and no one was directly injured. The event at Three Mile Island occurred from faulty instrumentation that gave erroneous readings for the reactor vessel environment. A series of equipment failures and human errors along with inadequate instrumentation allowed the reactor core to be compromised and go into a partial melt. The radioactive water that was released from the core was confined within the containment building and very little radiation was released. In the Three Mile Island incident, the safety devices worked as planned and prevented any serious injury. This accident resulted in improved procedures, instrumentation, and safety systems being implemented.

The Three Mile Island Unit 2 core has been cleaned up and the radioactive deposit stored. The Three Mile Island Unit 1 is still operating with a clean record.

Status of Nuclear

There are 104 nuclear plants in the U.S. that provide about 20% of our total power. Over half our power is generated by coal burning plants.

Nuclear power generation emits gases that harm the ozone layer. Radioactive waste and gases produced by nuclear power plants have been blamed for cancer, birth defects and immune system disorders in surrounding communities.

Before the 1979 partial meltdown at the Three Mile Island nuclear power plant in Pennsylvania, many electric utilities bought the nuclear option on the premise that it would be too cheap to meter.

In 1986, a disastrous accident occurred at the Chernobyl nuclear power plant in the former Soviet Union releasing radiation in the surrounding area. Public opposition to nuclear power grew and in the U.S., 117 nuclear reactors were canceled. These cancellations outnumbered the country's 103 operating reactors. One or two plants came online in the mid-1990s and no others were scheduled.

Nuclear Upswing

When the energy crisis occurred in California, many began calling for increases in energy production and that nuclear power be considered.

The Nuclear Energy Institute (NEI) is the industry's main lobbying group and publishes data about the cost of nuclear power. Its studies show that nuclear production costs are lower than other central power sources, including coal. The figures from the NEI are 1.83 cents per kilowatt hours for nuclear, 2.07 cents for coal and 3.52 cents for natural gas. These figures are the operating costs of running the reactors.

Another viewpoint is that nuclear energy is expensive, damaging to the environment and dangerous to human health. When you include the cost of building plants and dealing with nuclear waste, nuclear power is far more costly. Some of the costs of nuclear power have been paid by ratepayers. The capital costs of building nuclear plants has increased greatly over the decades. Much of this has been due to increased regulations pushing some plants up to the $10 billion range with the many modifications required.

The costs of dealing with the reactors' radioactive waste are estimated at $58 billion according to the Department of Energy. The costs of decommissioning, the tear down and clean up of old nuclear plants is also high.

Decommissioning the Yankee Rowe plant in Massachusetts, which is about one-seventh the size of the largest nuclear reactor now operating, is expected to cost almost $500 million according to the Nuclear Information & Resource Service.

Nuclear plant utilities are protected from nuclear accidents under the federal Price-Anderson Act, which was passed in 1957. A utility's liability for an accident is limited to $7 billion. The current estimate of Chernobyl's costs exceeds $350 billion.

Nuclear Waste

More than $6 billion has been spent on high-level waste disposal. Spent fuel remains deadly for at least tens of thousands of years. In order to keep it isolated from the environment, nuclear waste is to be buried deep underground. Nevada's Yucca Mountain, is currently under consideration. Many in the nuclear industry believe that the Clinton administration blocked action on this site to gain support in this area.

Nuclear disasters are also excluded from ordinary insurance policies such as earthquakes and other natural disasters. The *Houston Chronicle* called Yucca Mountain probably the most studied area in history. Claims by the federal government that the environmental effects of the repository will be small and have essentially no adverse impact on public health and safety have been challenged. It remains to be seen if there is the political will to go ahead with the site.

Yucca Mountain is in an active seismic area, and growing evidence indicates that it may not safely contain the 70,000 metric tons of spent fuel intended for 10,000 years of undisturbed storage. The ground around the mountain is expanding at a faster rate than DOE had originally reported which indicates an increased risk of earthquakes and volcanoes.

The mountain also has rain water flows to the aquifer under the mountain, picking up minerals that are corrosive to the nickel alloy that DOE plans to use for the waste containers.

Clean Nuclear

For a decade, the nuclear industry has been promoting nuclear as a clean source of energy that, unlike fossil fuels, produces no greenhouse gases or air pollution. Advocates like to claim nuclear power is environmentally friendly because it does not contribute to global warming the way fossil fuels do. Unlike coal, natural gas and oil-fired power plants, nuclear plants are free not only of carbon emissions but also of other noxious gases like sulfur dioxide, mercury and nitrogen oxide that have made fossil-fuel burning plants the biggest source of air pollution in the United States.

While nuclear energy does not produce as much CO_2 or other greenhouse gases as fossil power, it's inaccurate to call nuclear technology CO_2 free. A large amount of electric power is used to enrich the uranium fuel, and the plants that manufacture the fuel in the U.S. are powered with coal. The total impact on the greenhouse gases is not large globally but it is significant compared to the impact of the processes that create renewable energy.

Uranium production does have a notable impact on ozone depletion. The Environmental Protection Agency's (EPA) Toxic Release Inventory showed that in 1999, the nation's two commercial nuclear fuel-manufacturing plants released 88% of the ozone-depleting chemical CFC-11 by industrial sources in the U.S. and 14% of the discharges in the whole world.

References

Braun, Harry, *The Phoenix Project: An Energy Transition to Renewable Resources*, Research Analysts: Phoenix, AZ, 1990.

Cothran, Helen, Book Editor, *Global Resources: Opposing Viewpoints*, Greenhaven Press,: San Diego, CA, 2003.

Romm, Joseph J., *The Hype About Hydrogen*, Island Press: Washington, Covelo, London, 2004.

INDEX